T0137876

Springer Series in Adaptive Environments

The Springer Series in Adaptive Environments presents cutting-edge research around spatial constructs and systems that are specifically designed to be adaptive to their surroundings and to their inhabitants. The creation and understanding of such adaptive Environments spans the expertise of multiple disciplines, from architecture to design, from materials to urban research, from wearable technologies to robotics, from data mining to machine learning and from sociology to psychology. The focus is on the interaction between human and non-human agents, with people being both the drivers and the recipients of adaptivity embedded into environments. There is emphasis on design, from the inception to the development and to the operation of adaptive environments, while taking into account that digital technologies underpin the experimental and everyday implementations in this area.

Books in the series will be authored or edited volumes addressing a wide variety of topics related to Adaptive Environments (AEs) including:

- Interaction and inhabitation of adaptive environments
- Design to production and operation of adaptive environments
- Wearable and pervasive sensing
- Data acquisition, data mining, machine learning
- Human-robot collaborative interaction
- User interfaces for adaptive and self-learning environments
- Materials and adaptivity
- Methods for studying adaptive environments
- The history of adaptivity
- Biological and emergent buildings and cities

Philippe Morel · Henriette Bier
Editors

Disruptive Technologies: The Convergence of New Paradigms in Architecture

Editors
Philippe Morel
The Bartlett School of Architecture
University College London
London, UK

Henriette Bier
Delft University of Technology
Delft, Zuid-Holland, The Netherlands

ISSN 2522-5529 ISSN 2522-5537 (electronic)
Springer Series in Adaptive Environments
ISBN 978-3-031-14162-1 ISBN 978-3-031-14160-7 (eBook)
https://doi.org/10.1007/978-3-031-14160-7

This Springer imprint is published by the registered company Springer Nature Switzerland AG
The registered company address is: Gewerbestrasse 11, 6330 Cham, Switzerland

Preface

Reflections on Disruptive Technologies and the Convergence of New Paradigms in Architecture

Architects and scholars dealing with the evolution of design and building technologies (software and hardware) know that the disruptive nature of information technologies is not to be underestimated. Computation, i.e., raw computational power and abstract mathematical models that lead to new forms of design intelligence, is finally associated with novel forms of building production and operational intelligence. The convergence of virtual and material systems has opened the path to, as Chris Weil named it (2007), a 'phygital' world where the physical and digital are highly intertwined and connected. This convergence, as shown by authors such as Gerber and Ibanez (2014), Bier (2018), and Morel (2006), has immense consequences. From new sustainable energy generation and management principles to circularity and mobility concepts that feed into novel urbanistic and architectural approaches, they all take advantage of various degrees of machinic intelligence and are fundamentally changing the nature of physically built environments, their construction, and use.

Hence, contributors to this volume inquire if the use of smart manufacturing processes, for instance, distributed 3D printing with novel materials, or sensor-actuators embedded into production processes and buildings, imbue both construction and built environments with adequate degrees of intelligence. They ask if new geometries and topological explorations lead, indeed, to novel concepts of adaptive architecture and if the data-driven urban design or agent-based semiology delivers appropriate new forms of user-oriented public spaces and cyber-urban integration. They question if disruptive pedagogies for architecture or strategies for, according to Bhooshan et al., 'democratizing tectonism' create the basis for responsible design or should a radical shift in how architects engage with the discipline of architecture be considered. Answers to these questions are based on careful analyses of the evolution of technologies and critical reflection on the ongoing 'democratizing tectonism' offering a better insight into current disruptions in architecture.

While architectural knowledge is partly embedded in algorithms, leading directly or indirectly to the necessity of a machine-readable architectural approach, intelligence in architecture and building processes is increasingly achieved through human-robot/computer collaboration. The convergence of multiple technologies such as real-time analytics, machine learning, sensor-actuator networks, and automation (including building automation) facilitates the establishment of bio-cyber-physical feedback loops (Fass and Gechter 2016) that involve machinic intelligence ranging from basic intelligence level operating with if-then-else constructs to higher levels relying on machine learning. These facilitate the interaction between human and non-human agents engaged in the co-creation of adaptive and intelligent environments.

This volume investigates the hybrid human and non-human co-creation of adaptive environments with a strong emphasis on architecture and the data-driven shift within the epistemological framework of architecture. The contributions are organized into sections which address various social, economic, and scientific dimensions of the concepts, tools, technologies, and methodologies that led to new design paradigms and imperatives of the twenty-first century. Topics of machinic intelligence are addressed as well as topics relating to challenges and opportunities of new cyber-physical systems.

Given the explosion in access to data, this volume provides a historical context and a contemporary view of what it means to design with data, advanced software, and hardware. It establishes a meaningful dialogue between analysing and deriving performance in architecture with data, and producing or operating architecture with data, while also providing a critical reflection on limitations and misconceptions that need to be overcome in the discipline of architecture. In doing so, the volume addresses what is seen as a deep conceptual shift in science and design methodologies developed with the increasing access to machinic intelligence, i.e., artificial intelligence (AI). Specific cases and discussions on cyber-physical systems, human-machine interaction, levels of autonomy and interactivity, digital twins, and goal-driven approaches to address the imperatives of the twenty-first century are included. All in all, the presented works address the question of how digital techniques, simulations, and robotic production and operation represent a revolution for those involved in the design of new environments: designers, engineers, builders, and users. The three parts address (1) *Robotics and AI in Architecture*; (2) *Architectural Intelligence, Machine and Human Learning*; and (3) *Cyber-Urban Integration, Tectonism, and Disruptions*.

Part 1: Robotics and AI in Architecture

Advanced building technologies, robotics, and cyber-physical systems embedded in production processes and buildings fundamentally change today's architecture. The most important assumption explored in this section is that intelligence in architecture is neither located in the human nor in the non-human, i.e., software and hardware agents, but at the interface between them. Hence, this part presents the implications of this assumption, by reflecting on how robotic systems impact architecture due to the convergence of multiple technologies such as artificial intelligence (AI), large-scale machine-to-machine and human-to-machine communication (M2M and H2M), and the Internet of Things (IoT). These implications are explored and presented in relation to historical and theoretical interpretations and current manifestations by presenting ongoing research implemented at institutions such as McGill and Cornell Universities in North America, Technical University Delft in Europe, and the Chinese University of Hong Kong in Asia.

Theodora Vardouli is drawing from early research on 'responsive environments' that looks at topological ideas as both metaphors and operative artefacts for architectural adaptability. Her chapter gives an overview of topology's status in post-war mathematical and architectural cultures, including the Architecture Machine Group's (Negroponte 1973) efforts to produce computationally enhanced 'soft architectures' that co-evolve with their occupants using graph theory. Her argument is that both metaphors and operative artefacts help to historicize an imagination of design as fluid, soft, and malleable, while also foregrounding frictions with the discrete, symbolic logics of digital electronic computers—frictions that have practical and theoretical implications on contemporary perspectives on adaptive environments.

Practical implications on contemporary perspectives are presented by Yixiao Wang and Keith Green by reporting on user preferences for various interaction modes from pushbuttons to AI when interacting with robot surfaces—malleable, adaptive, physical surfaces that spatially reconfigure interior spaces within the built environment. They argue that with global mass-urbanization, the utility of robot surfaces in reconfiguring compact space into 'many spaces' is supporting and augmenting human activity. The question of the interaction between humans and such space is explored with users in a lab study at Cornell. It identifies preferences as split between AI- and user-controlled interactions because of the contexts of different scenarios and the complexity, accuracy, discretness, and feedback speed of different interaction modes.

AI- and user-controlled interactions are explored by Henriette Bier et al. via Design to Robotic Production and Operation (D2RP&O) processes that link computational design to materialization and operation of responsive building components. These processes are presented in a case study involving the development of urban interventions that activate residual spaces by introducing diversification of flora and fauna and by engaging neighbours and passers-by in 'caring' for the new species that are colonizing those spaces.

Urban spaces and their use are the focus of investigation for Jeroen van Ameijde as well. He explores how computational tools for site analysis and monitoring enable data-driven urban place studies that connect to generative strategies for public spaces and environments at various scales. He argues that today's 'smart city' initiatives seem to be contemporary interpretations of Negroponte's vision of computational processes that are open to participation and presents a series of theoretical and procedural experiments conducted through academic research and education, involving user-driven generative design processes in the spirit of 'The Architecture Machine'.

While all chapters in terms of content acknowledge that the advent of ubiquitous computing, and the embedding of sensing and actuating technologies in buildings and building processes, open up new opportunities for design, production, and operation in architecture and the built environment, approaches differ ranging from theoretical to more applied. All involve at some level robotics, AI, and/or user-controlled interactions. It is generally acknowledged that the design of physical environments incorporating sensor-actuators concerns (a) physical environment, (b) information flows and processes as well as (c) H/M2M communication. The challenges to integrating the design of interactions with the design of physical environments are addressed by establishing feedback loops and by relying on the understanding that the physical environment consists of building components that are cyber-physical in nature and their design and production are informed by material, structural, functional, environmental, and operational considerations. While robotic systems can significantly contribute to improving material-, energy-, and process efficiency, as well as the structural, environmental, functional, and operational performance of buildings and building processes (Sawhney et al. 2020), it appears that a reasonable number of tasks cannot be completely automated. Hence, robots will not completely replace humans but rather support them by firstly taking over repetitive and/or heavy tasks. Within this scenario, the human role will be mainly focused on envisioning new forms of physical environments and advancing novel means for their construction and operation as well as supervising and intervening when the non-human agents require assistance. This implies that cyber-physical systems integrated into buildings and building processes will increasingly share agency with humans in the use of means of production and space.

Part 2: Architectural Intelligence, Machine and Human Learning

The main concern of this part is the association of the *informational*, the *communicational*, and the *computational* within current technologies. While it was referred to as postmodern (Lyotard 1979), from the 1960s until around the end of the 1990s, the information society was still often treated as a modern society made of separate domains, in which the service sector would have taken precedence over the manufacturing, material goods and services circulating according to a new logic of networks (Castells 1998) which also governed human relations, relations of power, or simple relations of friendship. In line with its original name given by its creator Shannon (1948)—a (mathematical) "theory of communication"—this information society was also named the communication society (1948). In architecture, this communicational aspect was confirmed by Venturi (1996) who affirmed that *"modern architecture is about space, postmodern architecture is about communication"* (Venturi 1996). Although no one can deny the current importance of the communication phenomenon, it would nevertheless be wrong to limit the world to it. The world, in fact, is at the same time *informational, communicational,* and *computational*, and it is indeed this triple nature that is an urgent question today. Matter has not disappeared; it is as crucial as in all previous eras, but it is nowadays dominated by information. It is either a source of 'raw informational material,' or a vector of information, or both. It is also a support of computation that can be programmed as desired according to diverse models of computation. The consequences of these transformations, or rather, the massive consequences of these transformations within the architectural discipline are discussed in this section dedicated to architectural intelligence, human and machine learning.

In the first of four chapters, entitled "Architectural Knowledge and Learning Algorithms", Roberto Bottazzi (The Bartlett School of Architecture—University College London) elaborates on the new conditions that govern architectural knowledge in the age of machine learning algorithms. The interactions between humans and the ever more complex and foreign field of these algorithms are explored in relation to the *"complexity and cultural richness incorporated in the thought automation project"*. Bottazzi identifies the necessity to go beyond the mere understanding of the technical (algorithmic) aspect of the learning problem alone since *"learning algorithms pose more complex and conceptual challenges as they suggest a radical reorganization of space and scale."* These algorithms reorganize not only space but also the representation of urban complexity provided by data, while by organizing these data, they, in turn, provide new representations, certainly intelligible, but partial.

In the second chapter—"On Legibility: Machine Readable Architecture"—Andrew Witt (Harvard Graduate School of Design) deals with the concept of architectural and computational readability, encoding, and visual language in architecture. Witt proposes *"three related frames through which to interpret the entangled practices of architectural and machine readability. The first is a capsule chronology of machine readability, from its roots in tabular statistical datasets in the nineteenth century to its convergence with AI and machine learning today [...]. The second is an examination of the concept of architectural readability as it evolved complimentarily in the 1970s [...]. The third explores the intersections of the first two through the presentation of two design projects that use machine vision and machine readability [...]."* According to Witt, new ways of reading and generating architectural forms are needed, through the concept of machinic reading. *"From projects that morphologically catalogue the world's billion buildings to the application of shape classification for radical waste reuse,"* this *"machinic reading is transforming the roles and products of design."*

In the third chapter of this part, entitled "Where is Reality? Can You Show It to Me? Constructing Artificial Agency", Theodore Spyropoulos (Architectural Association School of Architecture) goes back to one of the founding theories of the information society, i.e., cybernetics: the 'first cybernetics' but also the one called 'second cybernetics' (or 'higher order cybernetics'). Summoning the English psychiatrist William Ross Ashby in his book titled *An Introduction to Cybernetics* (1956), Th. Spyropoulos insists on the fact that while cybernetics is a *'theory of machines'*, it deals not with objects but with *'ways of behaviour'*. *"It does not ask 'what is this thing?' but 'what does it do?'"*. Following on from this, and from the reading by

Johnston (2008) of Ashby's cybernetics, Spyropoulos notes that the real object of this theory is the *'domain of all possible machines.'* Whether some of these machines were not made by man or by nature is a secondary question. What cybernetics truly offers is a *"framework on which all individual machines may be ordered, related and understood."* For Spyropoulos, what matters is not machines as such, as informational machines, but their behaviour and communication potential within this framework common to machines and humans. Starting from questions about (massive) communication, Spyropoulos also poses the ontological question of the nature of reality. Where is it? In the minds of humans or in the memories of machines?

In the last chapter—"From Disruptions in Architectural Pedagogy to Disruptive Pedagogies for Architecture"—Sevgi Türkkan (Istanbul Technical University) is concerned with architectural education, a question 'a fortiori', in view of the current radical transformations. Her chapter is a pedagogically oriented reflection addressing the techno-cultural-pedagogic shift in architectural education. A manifesto for forms of intelligence, labour, creativity, and reorganization of space offered for more relevant architectural learning, it is calling for radical changes in the pedagogic agenda, thanks to recent advancements in digital knowledge, big data availability, and open-source AI tools. It challenges in a very concrete manner the *"mainstream and 'ordinary' architecture school, its educational concepts, curriculum, pedagogic rituals, values, and the disciplinary ethos that lies underneath."* As Türkkan mentions, the aim of this chapter is *"to outline trajectories for this agenda, by raising a series of questions regarding architectural learning and the role of institutions in the 21st century"*.

Part 3: Cyber-Urban Integration, Tectonism, and Disruptions

After addressing the issues of robotics, human-robot interaction, artificial and architectural intelligence, and machine and human learning, the third and last section of this volume is dedicated to the broader context of urban design in the age of omnipotent cyberspace, technological appropriation, and disruptive innovation beyond architecture. When looking at recent changes in architecture, in the last 50 years, between 1972 and 2022, the characteristic that strikes most is the gap between the architecture commonly built in 1972—including its modes of practice that would be called today 'business models'—and that built in the last decade. Although 50 years is a very short time in the history of architecture, and technology, 1972 seems just as far today as the steam engine. Indeed, the debates of an era that now embodies the birth of postmodernism, or at least a form of culturalist postmodernism, seem to today naïve, 'arty', and self-centred on the intellectual elite that produced it. These debates also seem reductionist, or at least very much out of step with the radicality of the transformations at work in the organization and planning of business at the global level, as for instance: the computerization of trade and markets with the computerization of the National Association of Securities Dealers Automated Quotations (NASDAQ) in 1971, the explosion of container-based logistics, urban hyper-growth, the explosion of tourism, the end of the Bretton-Woods agreements from August 1971, etc. While from the end of the 1960s onwards certain architects (e.g., John Negroponte) and technologists tried, *by technological means*, to take better account of urban realities and the needs and wishes of the inhabitants, what will remain overall from this era will be social experiments in direct participation that were quickly rendered obsolete by the complexity and slowness of the decision-making processes, confronted by the speed and power of the market. Today, in a post-Internet era that seems already formidably accomplished, even though it is only an embryonic state of a new civilization, there are new calls for an architecture that is fundamentally connected to reality—but to *the whole of reality*, not to its formal, stylistic, aesthetic, social or economic aspects taken individually. Some of these calls come from the authors of the chapters in this section "Cyber-Urban Integration, Tectonism, and Disruptions". Vishu Bhooshan, Henry David Louth, and Shajay Bhooshan advocate the need for a new use of advanced technologies and a new form of provision of both the tools and the results of their use by architects; Philippe Morel reminds us of the need for new forms of practice and, to this end, for a better knowledge of the mechanisms of innovation. As for Patrik Schumacher, he is undoubtedly the practitioner and theorist whose work has the widest visibility, audience, and impact. While 'parametricism' has been perceived as a new attempt to restore a style, which its author has defended, arguing that only a style has the power to transmit a new set of values, no attentive reader can deny that the theoretical richness of this concept goes far beyond this issue. Hence, the first of the three chapters by Schumacher (Zaha Hadid Architects, Architectural Association), entitled "Cyber-Urban Integration", represents a further development of Schumacher's thinking. It speculates on the current integration of the digital and the physical within new *'cyber-urban'* environments. According to the author, *"after 30 years of theoretical speculation and advances in gaming and entertainment, the internet is finally on the way to transforming into cyberspace. The magazine as a guiding analogy for the web is being overtaken by the analogy of the city. Architects take over from graphic designers. The premise for the plausibility of this takeover and expansion of architecture's competency is that all design, including architecture, is communicative framing. The thesis of this paper is that in this age of soaring web-based telecommunication, the space of social communication must be designed simultaneously as a physical and virtual realm, as a cyber-urban space, seamlessly integrating physically immediate and digitally mediated communicative interactions, constituting a new augmented mixed reality."* In his chapter, Schumacher elaborates on the nature of *"architecture's core competency"* through what he calls *"the four architectural projects"*. He shows how these projects are dependent on a new industrial system, a new *"pro-active Intelligent Environments"* and an agent-based parametric semiology that, according to him, should be expended to realize the full potential of a finally mature cyberspace within the discipline of architecture and beyond. Such cyberspace represents, according to Michael Benedikt whose 1991 book *Cyberspace: First Steps* is discussed by Schumacher, *"a new stage, a new and irresistible development in the elaboration of human culture"* (Benedikt 1991).

The second chapter—"Democratising Tectonism: A High-Performance Technological Basis for Engaging and Responsible Design", Online and On-land—by Vishu Bhooshan, Henry David Louth, and Shajay Bhooshan (Zaha Hadid Architects, Architectural Association) deals with the possibility of such democratization through the concept of *"Spatial Technology Stack (STS)"* that unifies Architectural Geometry and game-tech. According to the authors, such an STS could *"robustly support the synthesis of high-performance shapes including structurally optimized geometry and its processing for robotic and digital fabrication (RDF), and the creation of environments that deliver novel, engaging and productive spatial user experiences both in the physical and virtual instantiations of architecture"*. Contrary to *"misaligned building information modelling technologies"*, the STS could finally provide *"an alternative high-performance technological basis for engaging and responsible design, both online and on-land"*, within the context of a new *"cultural production view of architecture, spatial user-experience (UX) design, and end-user ergonomics"*.

The third and last chapter of the part, by Philippe Morel (Associate Professor at UCL Bartlett and ENSA Paris-Malaquais, initiator and founding CEO of XtreeE), entitled "Why Disruptive Business Models are Inseparable from Disruptive Technologies", goes back to the importance of novel business models in today's technological explosion. It addresses the relationship between business models and disruptive technologies as a counterpoint to the general theme of this volume *Disruptive Technologies: The Convergence of New Paradigms in Architecture*. While discourse on disruptive technologies commonly insists on the technologies themselves, most often from the point of view of their technical operativity or from an epistemological perspective, a closer look at the reality of techno-capitalist societies reveals the crucial importance of how technologies are inserted into the global economic market. This insertion obviously impacts the technological appropriation, but maybe more importantly the technological evolution itself, including in architecture perceived here in a broad sense, from the conception to the maintenance of projects after delivery. By looking at a few arguments about the nature of disruptive technology and innovation, including from the inventor of that very notion of disruptive innovation, the final chapter demonstrates how different the current time is from everything that preceded it. Indeed, while business models in architecture have rarely ever changed until the beginning of the twenty-first century, new models might become one of the most important parameters of change in the post-Internet era, beyond mere technological change which is far too often the unique concern of architects.

In conclusion, adaptive and intelligent environments are experiencing exponential growth and becoming a pervasive component of the design, construction, and operational explorations in architecture. Through the lenses of concepts, data, tools, methods, and imperatives, contributors argue for a contemporary understanding of the deep computational, cyber-physical revolution in architecture. Issues of the disappearance of aesthetics and yet the emergence of other aesthetics and criteria brought about by data are discussed and a discourse on disruption and novel models of practice and innovation is raised, and a discussion on the spectrum of paradigms and shifts is expanded.

London, UK Philippe Morel
Delft, The Netherlands Henriette Bier

Acknowledgements The editors of this volume would like to warmly thank the series editors Holger Schnädelbach and Kristof van Laerhoven for their generous invitation, Springer editors Helen Desmond and Barbara Amorese for their continuous support and patience, and the authors for their deep engagement with the topic and the quality of their respective contributions. They also want to thank David Gerber and point out that this book has profited from his highly generous input.

References

Benedikt M (ed) (1991) Cyberspace: first steps. MIT Press, Cambridge MA
Bier H (2018) Robotic building. Springer. 10.1007/978-3-319-70866-9
Castells M (1996 to 1998) The rise of the network society, the information age: economy, society and culture, vol 1. Oxford, Malden, Blackwell, UK, 1996. The power of identity, the information age: economy, society and culture, vol 2. Oxford, Malden, Blackwell, UK, 1997. End of millennium, the information age: economy, society and culture, vol 3. Oxford, Malden, Blackwell, UK, 1998
Fass D, Gechter F (2016) Towards a theory for bio−cyber physical systems modelling. arXiv: 1601.06962v1 [q-bio.QM]
Gerber D, Ibanez M (2014) Paradigms in computing: making, machines, and models for design agency in architecture. eVolo. ISBN-13: 978-1938740091
Johnston J, John (2008) The allure of machinic life: cybernetics, artificial life, and the new AI. The MIT Press
Lyotard J-F (1979) La Condition postmoderne. Les éditions de minuit, Paris
Morel P (2006) Architecture et Convergence Technologique (architecture and technological convergence), lecture at IFA-Institut Français d'Architecture (French Institute of Architecture), Paris, France, February 2003; and From e-Factory to Ambient Factory (Or What Comes After Research?). In: Oosterhuis K, Feireiss L (eds) Game set and match II. On computer games, advanced geometries, and digital technologies. Episode Publishers
Negroponte N (1973) The architecture machine—towards a more human environment. MIT Press, 1970
Sawhney B, Riley M, Irizarry J (2020) Construction 4.0—an innovation platform for the built environment. Routledge. 10.1201/9780429398100
Shannon CE (1948) A mathematical theory of communication. Bell Syst Tech J 27(3):379–423
Venturi R (1996) Iconography and electronics upon a generic architecture: a view from the drafting room. MIT Press

Contents

About the Editors

Philippe Morel is an architect, theorist and entrepreneur, co-founder of EZCT Architecture and Design Research (2000) and initiator and founding CEO of the large-scale 3D-printing corporation XtreeE (2015). He currently teaches as an Associate Professor at the École nationale supérieure d'architecture Paris-Malaquais, where he headed the Digital Knowledge department he co-founded with Prof. Girard, and at UCL Bartlett where he is the Architectural Computation MSc/MRes (BPro) program director. Before teaching at the Bartlett, he was a seminar and studio Professor at the Berlage Institute in Rotterdam, and a history and theory seminar and AADRL tutor at the Architectural Association in London. In 2017, he co-edited the book Computational Politics and Architecture: From Digital Philosophy to the End of Work (Editions ENSAPM). In February 2007, he curated the exhibition Architecture Beyond Forms: The Computational Turn at the Maison de l'architecture et de la ville PACA in Marseille. Explicitly departing from Eisenman Ph.D.'s dissertation from 1963—The Formal Basis of Modern Architecture—the exhibition addressed both historically and theoretically the current linguistic and computational turns in architectural design. Philippe Morel has published more than 40 essays, lectured at many universities around the world and presented his work in numerous exhibitions. Philippe Morel's work and the one of his office are part of private and public collections, including FRAC Centre and Centre Pompidou permanent collections.

Henriette Bier After graduating in architecture (1998) from the University of Karlsruhe in Germany, she has worked with Morphosis (1999–2001) on internationally relevant projects in the US and Europe. She has taught computational design (2002–2003) at universities in Austria, Germany, Belgium and the Netherlands, and since 2004 she mainly teaches and researches at Technical University Delft (TUD) with a focus on robotics in architecture. She initiated and coordinated (2005–2006) the workshop and lecture series on Digital Design and Fabrication with invited guests from MIT and ETHZ and finalized (2008) her Ph.D. on System-embedded Intelligence in Architecture. She coordinated EU-funded projects E-Archidoct and F2F Continuum (2007–2010), led NL-funded projects Scalable Porosity and Adaptive Stiffness (2015–2018), and ESA-funded project Rhizome (2021–2022). During

2017–2019, she has been appointed professor at Dessau Institute of Architecture and a 2020 visiting researcher at PoliMi. The results of her research are published in books, journals and conference proceedings and she regularly lectures and leads workshops internationally.

Part I
Robotics and AI in Architecture

Chapter 1
Implications of Robotics and AI in Architecture

Henriette Bier

Robotic systems are increasingly incorporated into building processes and buildings. The question for the future is thus not if but how robotic systems will be integrated into architecture and the built environment. Such systems have a major impact due to the convergence of multiple technologies such as artificial intelligence (AI), large-scale machine-to-machine and human-to-machine communication (M2M and H2M), and the Internet of Things (IoT). Implications are explored and presented in this section in relationship to historical and theoretical interpretations and current manifestations by presenting ongoing research implemented at institutions such as McGill and Cornell Universities from North America, Technical University Delft from Europe, and the Chinese University of Hong Kong from Asia.

Theodora Vardouli is drawing from early research on 'responsive environments' that is looking at topological ideas as both metaphors and operative artifacts for architectural adaptability. The chapter gives an overview of topology's status in post-war mathematical and architectural cultures, including the Architecture Machine Group's efforts to produce computationally enhanced 'soft architectures' that co-evolve with their occupants using graph theory. Her argument is that both, metaphors and operative artifacts, help to historicize an imagination of design as fluid, soft, and malleable, while also foregrounding frictions with the discrete, symbolic logics of digital electronic computers—frictions that have practical and theoretical implications on contemporary perspectives on adaptive environments.

Practical implications on contemporary perspectives are presented by Yixiao Wang and Keith Green by reporting on user preferences for various interaction modes from pushbuttons to AI when interacting with robot surfaces—malleable, adaptive,

H. Bier (✉)
Faculty of Architecture and the Built Environment, Delft University of Technology (TU Delft), Delft, Netherlands
e-mail: H.H.Bier@tudelft.nl

P. Morel and H. Bier (eds.), *Disruptive Technologies: The Convergence of New Paradigms in Architecture*, Springer Series in Adaptive Environments,
https://doi.org/10.1007/978-3-031-14160-7_1

physical surfaces that spatially reconfigure interior spaces within the built environment. They argue that with global mass-urbanization, the utility of robot surfaces in reconfiguring compact space into 'many spaces' is supporting and augmenting human activity. The question of the interaction between human and such space is explored with users in a lab study at Cornell. It identifies preferences as split between AI- and user-controlled interactions because of the contexts of different scenarios and the complexity, accuracy, discretness, and feedback speed of different interaction modes.

AI- and user-controlled interactions are explored by Henriette Bier et al. via Design to Robotic Production and Operation (D2RPO) processes that link computational design to the materialization and operation of responsive building components. These processes are presented in a case study involving the development of urban interventions that activate residual spaces by introducing diversification of flora and fauna and by engaging neighbors and passers-by in 'caring' for the new species that are colonizing those spaces.

Urban spaces and their use are the focus of investigation for Jeroen van Ameijde as well. He explores how computational tools for site analysis and monitoring enable data-driven urban place studies that connect to generative strategies for public spaces and environments at various scales. He argues that today's 'smart city' initiatives seem to be contemporary interpretations of Negroponte's vision of computational processes that are open to participation and presents a series of theoretical and procedural experiments conducted through academic research and education, involving user-driven generative design processes in the spirit of 'The Architecture Machine' (Negroponte 1970).

While all chapters in terms of content acknowledge that the advent of ubiquitous computing, and the embedding of sensing and actuating technologies in buildings and building processes, open up new scenarios for design, production, and operation in architecture and the built environment, approaches differ ranging from theoretical to more applied. All are involved at some level in robotics, AI- and/or user-controlled interactions. It is generally acknowledged that the design of physical environments incorporating sensor-actuators are concerning (1) physical environment, (2) information flows and processes as well as (3) H/M2M communication. The challenges to integrating the design of interactions with the design of physical environments are addressed by establishing feedback loops and by relying on the understanding that the physical environment is consisting of building components that are cyber-physical in nature, and their design and production are informed by material, structural, functional, environmental, and operational considerations (Fig. 1.1).

While robotic systems can significantly contribute to improving material-, energy-, and process efficiency, as well as the structural, environmental, functional, and operational performance of buildings and building processes (Inter al; Bier 2018; Sawhney et al. 2020), it is to be expected that data-driven automation involving AI will cover 50% of all tasks, whereas 45% will rely on human–robot interaction (HRI) and 5% will require human intervention (Inter al). Hence, humans and robots will operate building processes and buildings side by side.

Fig. 1.1 Human-assisted robotic assembly of non-uniform linear elements implemented with PhD and MSc students © RB lab, TUD and UASA

Considering the 50% of tasks that cannot be completely automated, robots will not replace humans but rather support them by taking over repetitive and/or heavy tasks. The human role will be mainly focused on envisioning new forms of physical environments and advancing novel means for their construction and operation as well as supervising and intervening when the non-human agents require assistance. This implies that cyber-physical systems integrated into buildings and building processes will increasingly share agency in the use of means of production and space.

References

Bier H (2018) Robotic building. Springer. https://doi.org/10.1007/978-3-319-70866-9

Inter al. McKinsey report. https://www.mckinsey.com/business-functions/mckinsey-digital/our-insights/winning-in--requires-a-focus-on-humans.

Negroponte N (1970) [1973] The architecture machine—Towards a more human environment. MIT Press

Sawhney B, Riley M, Irizarry J (2020) Construction 4.0—An innovation platform for the built environment. Routledge. https://doi.org/10.1201/9780429398100

Chapter 2
Bioptemes and Mechy Max Systems: Topological Imaginations of Adaptive Architecture

Theodora Vardouli

2.1 Introduction

In 1971, the video and new media journal *Radical Software* published excerpts of a manuscript and video interview by a Canadian psychiatrist, psychoanalyst, and cybernetician Warren Brodey, in which he urged for a new way of conceptualizing, and intervening in, "context" (Brodey 1971: 4). His polemic was founded upon two neologisms—"bioptemes" and "mechy max systems". Standing for "biological optimizing systems" and "mechanically maximizing systems", respectively, the two terms reflected opposing approaches to relations between an entity and its encompassing systems. In mechy max systems, all the elements of the system were fixed and stable, whereas in bioptemes the elements negotiated their boundaries in dynamic processes of mutual adaptation. Brodey lambasted designers and technologists for their long-standing preoccupation with mechy max-es which was preventing them from achieving the fluidity and adaptability of biological systems.

Contemporary debates on "adaptive architecture" echo a similar lament. In introducing the edited volume that came out of the 2013 conference *Alive: Advancements in Adaptive Architecture* at ETH Zurich, Manuel Kretzer critiqued contemporary architectural design for its functionalist tropes and continued commitment to standards and stabilities in the conception of people and buildings (Kretzer 2014: 17). Cutting against the grain of these disciplinary inertias, he argued, was a legacy of architectural experimentation with adaptive architecture that dated back to the 1960s and 1970s. Despite its arguable heterogeneity in terms of politics and approaches to technology, the group of architects that Kretzer invoked—including names like Yona

T. Vardouli (✉)
Peter Guo-hua Fu School of Architecture, McGill University, Montreal, Canada
e-mail: theodora.vardouli@mcgill.ca

P. Morel and H. Bier (eds.), *Disruptive Technologies: The Convergence of New Paradigms in Architecture*, Springer Series in Adaptive Environments,
https://doi.org/10.1007/978-3-031-14160-7_2

Friedman, Constant Nieuwenhuys, Archigram, Superstudio, and others—all imagined an architecture that was, like all things alive, capable of change and adaptation to various forces.

Along with renegotiating the roles and agential potentials of architecture's subjects (Busbea 2020), postwar engagements with responsiveness and adaptability were also animated by a wholesale reconsideration of design processes and their mediating techniques. This chapter focuses on architects' uptake of new mathematical and calculative techniques that allowed to visualize objects and spaces in flux and to devise new protocols for change, choice, and possibility. These calculative and representational techniques challenged the primacy of Euclidean geometry in conceptualizing and visualizing architectural form—a discourse that is common in current adaptive architecture debates. Kretzer, for instance, argued that "alive" architecture evades geometric representation. "We do not believe," he wrote, "that it is possible to do justice to today's architecture with Euclidean or analytical geometry any longer" (Kretzer 2014: 13). Non-Euclidean geometry, he continued, appears to be a better candidate, but is compromised by computer visualizations that tame it under "instrumental and analytical interpretation patterns" (Kretzer 2014: 13). In other words, an architecture that is truly "alive" transcends static representation and its geometric armature.

The chapter is based on the premise that despite its advocates' patent aversion to fixed architectural forms and images, adaptive architecture is undergirded by stable imagery of mathematical entities. Coming from the field of topology, "the science of properties of spaces and figures that remain unchanged under continuous deformations" (Epple 1998: 299), these entities served as a springboard for an architectural imagination of adaptability, evoking shapes in continuous transformation and new possibilities of spatial reasoning. Topological entities were at the same time instruments deployed for descriptive, analytical, and calculative purposes and cultural objects with aesthetic qualities and discursive functions.

As art historian Linda Darlymple Henderson has shown in her classic study of non-Euclidean geometry in twentieth-century art, as mathematical ideas begin to sip into culture, they become vested with epistemic, aesthetic, philosophical, and even spiritual meanings (Henderson 2013: 92). Returning to the example of Brodey's "bioptemes," it was the topological construct of the Klein bottle—with its ambiguity between inside and outside—that provided an image and a conceptual handle for the involutions between things and their environments (Fig. 2.1). As architectural historian Larry Busbea has compellingly described, the Klein bottle also served as a logo for a company that Brodey started with fellow cybernetician Avery Johnson and the basis of video artist Paul Ryan's landmark essay "Cybernetic Guerrilla Warfare" (Busbea 2020, 142; 164).

Acknowledging the conceptual, aesthetic, and rhetorical multivalence of mathematical entities and their concrete manifestations, to borrow historian of mathematics Alma Steingart's term (2015), I discuss three episodes in which topological entities figured as both metaphors and operative artifacts in envisioning architectural adaptability.

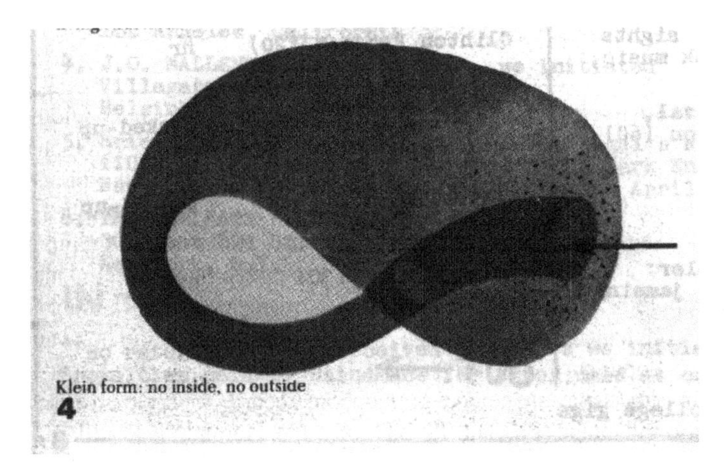

Fig. 2.1 Klein Form. *Source* Brodey (1971). Reproduced with permission of Davidson Gigliotti

My focus in this chapter is on topological entities that architects and designers enlisted to reason about issues pertaining to architectural "functions" and "programs," or put differently, what design and mathematics scholar Andrew Witt has termed "topologies of function" in juxtaposition to "topologies of form" (2022: 227; 259). I discuss graphs, networks, and other relational entities used to convey the *organization of spaces and functions* as opposed to properties of the building's shape itself. Although these entities' visual manifestations were not formally evocative in the same way as knots, Klein bottles, and other topological constructs, they supported equally vocal statements on form, its generation, and its status in the discipline of architecture (on responsiveness and topology see also Teyssot 2013). This chapter is structured around three episodes in which topological constructs were mobilized to support, and ultimately construct, divergent imaginations of adaptability.

Drawing from the etymological roots of "adaptability" from the latin *aptus,* which translates as "fit," I pay special attention to how similar mathematical techniques were mobilized, symbolically and technically, to support different approaches to "fitness." First, I discuss efforts to "fit" patterns of spatial organization with patterns of human activity. Then, I examine the notion of "misfit" as a structuring principle of a design process that promises a form that is "well-adapted" to its context. Finally, I move to "evolutionary fit" as the result of a "conversation" between human agents and an "intelligent" surround. Examining the topological constructs behind these expeditions toward adaptive architectures or architectures of "good fit" can help historicize the mathematical underpinnings of architecture as fluid, soft, and malleable, while also foregrounding frictions between these visions and the discrete logics of network topologies—frictions that have practical and theoretical implications for contemporary approaches to adaptive architecture.

2.2 Fitting Shapes—Activity Graphs

In the mid-1950s, architecture critic Reyner Banham was declaring topology's imminent victory over geometry. Discussing Alison and Peter Smithson's Golden Lane and Sheffield University competition entries, Banham wrote: "As a discipline of architecture topology has always been present in a subordinate and unrecognized way—qualities of penetration, circulation, inside and out, have always been important, but elementary Platonic geometry has been the master discipline. Now, [...] the roles are reversed, topology becomes the dominant and geometry becomes the subordinate discipline" (Banham 1996 [1955]: 14). Banham remarked on the profound impact of topology's dominance, in which "a brick is the same 'shape' as a billiard ball [...] and a teacup is the same 'shape' as a gramophone record" for notions of architectural form, beauty, and image (Banham 1996 [1955]: 14). Dominique Rouillard has documented the Smithsons' key role in the postwar wave of megastructural architecture (2004), which Larry Busbea has fittingly classified under the moniker "topologies" (2007). Less known, however, is the Smithsons' participation in the publication of one of the first textbooks on modern mathematics and architecture, which promoted, as the book's blurb noted, a "structural" understanding of the environment through the aid of modern mathematics.

As the anecdote goes, in the late 1960s Alison and Peter Smithson approached the Royal Institute of British Architects (RIBA) Library Committee with a grievance (March 2002: 30). In terms of their significance for architecture, the new mathematical varieties proliferating in the mid-twentieth century seemed to parallel the invention of linear perspective in the Renaissance. And yet that mathematics—to which the Smithsons had been exposed through their son's school textbooks—appeared obscure and inaccessible. To remedy the situation, the RIBA Library Committee invited Lionel March, the newly appointed Director of the Land Use Built Form Studies (LUBFS) Centre at the University of Cambridge Department of Architecture, to write a book on twentieth-century mathematics and architecture. March co-authored the book with LUBFS Centre member Philip Steadman, who had studied under Bryan Thwaites—a key figure in an impactful effort to modernize British school mathematics that came to be known as the "new math" (2021 [1971]).

Titled *The Geometry of Environment: An Introduction to Spatial Organization*, the book was organized based on "new math" topics. In the book's introduction, co-authors March and Steadman advocated for the benefits of the architectural reader's exposure to a new kind of geometry that did not have to do with measures and proportions, but instead with structures and relations—with topological ideas. The first seven chapters of the book, written by March, used architectural examples to illustrate mathematical concepts such as mappings, transformations, and symmetry groups. Chapters 8, 9, 10, 11, 12, 13, and 14, written by Steadman, presented applications of mathematics to architectural problems. The common characteristic of these "applied" chapters with titles such as "Electrical Networks and Mosaics of Rectangles," "Spatial Allocation Procedures," and "Networks Distances and Routes" was a consistent imagery of point and line diagrams: of graphs. These graphs could reveal hidden spatial patterns in famous architectural works—such as Frank Lloyd Wright's Devin House (March & Steadman 1974: 258), calculate new organizations of space congruent with patterns of human activity that they were also used to map, and provide new abstractions of the shape of buildings.

These graphs convey the first topological imagination of adaptive architecture that I wish to discuss here: an architecture that *fits*, is well adapted, to a structure of human activities. Activities, in this context, were conceptualized as tasks undertaken in discrete locations. Researchers in British government organizations, such as Ian Moore at the Offices Development Group of the Ministry of Public Building and Works, took "activities" as the fundamental unit for new methods of architectural programming that relied on empirical observations of occupants in various building types (Broadbent 1988 [1973]: 288). These were initially developed in the context of "scientific management" to tackle problems of representing workflows and calculating the optimal locations of bodies and equipment on the factory floor (Koopmans et al. 1957; Muther 1961; Stone 1963; Moseley 1963; Buffa et al. 1963; Whitehead & Eldars 1964). Data-collection methods had, by the early 1960s, made their way into the design of buildings such as hospitals (Theodore 2013).

Outputs of these efforts to collect information on the activities of hospital staff were representations such as "string diagrams" that visualized the activity pattern of healthcare staff on the hospital floor. Such diagrams, literally made from string, were soon after translated into data structures consisting of activity-location pairs and frequency of movements between them (Vardouli & Theodore 2021). This translation rendered string diagrams as graphs, and topological representations of activity patterns, which were subsequently used to generate a floor plan or a layout that best *fits* these activities. "Best" here was a synonym for "cost-effective" and referred to minimizing time spent walking between activities. A vexing question in this work was the stability of activities as well as their professed independence from the space in which they were observed in the first place. In *The Geometry of Environment*, March and Steadman referred to the choice of hospitals as the testing ground for space allocation work as "no accident" as "the pattern of hospital routine is perhaps more standardized and consistent than in other kinds of organization, and so variations in the trip pattern over a period of time might be less extreme" (1974: 332) (Fig. 2.2).

It seems unintuitive to discuss layout optimization and space allocation as a precursor to adaptive architecture. Fixed activity patterns, singular goals, and optimal layouts seem to have very little to do with adaptive architecture's embrace of time and change. And yet, the topological imagination of activity patterns and spatial organization was fundamental for imagining a floor plan geometry in flux. A telling example of such recasting is the chapter "Electrical Networks and Mosaics of Rectangles" in *The Geometry of Environment*, which was based on a working paper by Philip Steadman on the "Automatic Generation of Minimum-Standard House Plans." Building on generic plans and house shells published by the National Building Agency in the late 1960s, activity requirements by the Ministry of Housing and Local Government, and dimensional requirements by the Parker Morris Committee, Steadman sought to address the question of how to adopt a cost-effective set of (dimensional) standards

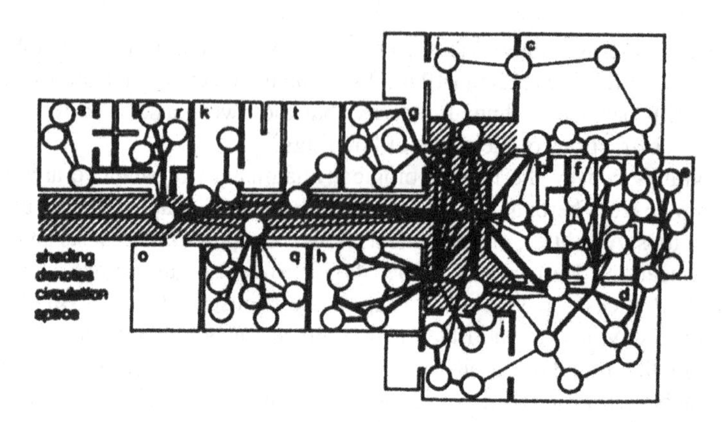

Fig. 2.2 Graph showing a nurse's daily movements around a hospital operating theater suite, redrawn for a *LUBFS Working Paper* from Whitehead and Eldars' 1964 article "An Approach to the Optimum Layout of Single-Storey Buildings." *Source* Tabor (1969). "Pedestrian Circulation in Offices." *LUBFS Working Paper* 17. Reproduced with permission of the Martin Center, University of Cambridge

for industrialized building construction, "without any significant reduction of choice in layout or design" (Steadman 1970: 21).

Steadman developed a method for enumerating all possible room-type adjacencies within a single floor and then packing rectangles of specific dimensions within a given outline (shell). The basis of this method was regarding the plan itself as a graph, with the walls or room boundaries being the graph's lines and their intersections being the graph's points. This graph, describing the physical elements of the architectural plan, was the "dual" of the adjacency graph—a topological description of types of rooms and their connections. Using this one-to-one translation between the functional and formal diagram of the floor plan, Steadman could calculate all possible layouts for specific activities and then apply further theorems and techniques from graph theory to pack specific room sizes within a given boundary (shell). This enumerative project, counting possible floor plans, was the basis of Steadman's work on "architectural morphology" or "the science of possible forms" that he continued at the Open University in Milton Keynes (Steadman 1983). The key to enumerating configurational possibilities was distinguishing between dimensional and "shape" properties, where "shape" was defined topologically as a general description for infinite dimensioned shapes (Steadman et al. 1991: 87).

In his 1983 book *Architectural Morphology*, Steadman made reference to—albeit critically for its mathematical imprecisions—a sketch of an unrealized machine for producing "menus" of floor plans (Steadman 1983: 144). The machine was Yona Friedman's FLATWRITER, a system developed for the 1970 Osaka World Expo as the implementation of a participatory design theory that he detailed in his 1971 book *Pour Une Architecture Scientifique* (trans. 1975 as *Toward a Scientific Architecture*). The FLATWRITER was the computer system that would allow the inhabitants of the *Ville spatialle*—a levitated three-dimensional infrastructure that Friedman envisioned as affording social and spatial mobility—to design the layouts of their homes and place them in the city. The FLATWRITER would use graphs and combinatorial methods to calculate all possible layouts for a given number of rooms selected by the inhabitant (Friedman 1971). The inhabitants would also monitor their daily habits by tracking how many times they entered a room or went to a specific location in the city. The FLATWRITER would then print a large book that would contain the "menu" of all possible layout choices along with a "warning," a measure of each floor plan's efficiency for the activity pattern tracked by the inhabitant—a surprising rhetorical dislocation of graphs from narratives of optimization to ones of choice, adaptive change, and self-expression.

2.3 Misfit Hierarchies—Trees, Cascades, and Networks

Space allocation research, along with its imagery of graphs and networks, proliferated in architectural research publications throughout the 1960s and 1970s. The rhetorical valency of these topological representations of work or dwelling ranged from prospects of an "automated architect" (Cross 1977) searching "spaces" of floor plan alternatives for those fitting a pattern of activities to visions of cities designed by their inhabitants after self-monitoring their daily movement patterns (Friedman 1971). A conspicuous *misfit* in this technical and discursive landscape was Cambridge University-trained architect and mathematician Christopher Alexander. Both an instigator and a critic of computers in architectural design, Alexander condemned space allocation processes for optimizing mathematical functions that relied, however, on faulty conceptual premises (Alexander 1967). Alexander, too, believed that computers could pave the path toward an architecture of "good fit" (1964: 15) as he called it—forms well adapted to their contexts. If space allocation was a "mechy max," relying on fixed entities of discrete tasks (activities) and discrete locations in space considered independently from each other and mapped onto each other through their representations as graphs, Alexander envisioned something closer to what Brodey would later refer to as a "biopteme." Instead of taking form (the physical, geometric characteristics of an artifact) and functional context as two separate systems, Alexander characterized them as an "ensemble" (Alexander 1964: 15).

Alexander's path toward an adaptive architecture was also paved with topological constructs: graphs, trees, semi-lattices, cascades, and networks. These entities worked not only as an invisible connective thread in Alexander's work but also as signposts of change in his theoretical approach to well-adapted form. Alexander was in fact among the first to bring topological entities such as graphs and trees into mainstream architectural parlance, in the book that has been described by its reviewers as a herald of worldwide efforts to devise rational methods for design (Montgomery 1970). The *Notes on the Synthesis of Form,* as the book was called, came out of Alexander's doctoral dissertation at Harvard.

The *Notes* presented a systematic method for generating designs not by striving to achieve goals—as was the approach in space allocation work—but by trying to eliminate what he called "misfits": all the ways in which a physical thing could *fail* to meet specific needs or other requirements. Alexander construed such failures as indications of a lack of adaptation between form and its functional context. In the *Notes,* Alexander used set and graph theory to calculate relationships among these misfits and use these relationships to group the misfits into a tree—a hierarchically ordered graph—comprised of simpler groups of misfits. The designer would "solve" these smaller groups through schematic drawings that Alexander called "diagrams" and compose these diagrams in the order prescribed by the tree to produce the full design. Because misfits were form-context relations, the tree ostensibly represented neither form nor context alone but their ensemble. "Good fit," Alexander wrote in the *Notes*, "is a desired property of this ensemble which relates to some particular division of the ensemble into form and context" (Alexander 1964: 16).

Statements about design as a process of form and context adaptations were inspired by the psychology of problem solving (Alexander cited German gestalt philosopher Karl Duncker and American cognitive psychologist George Miller). However, Alexander justified them not through recourse to contemporaneous debates on design problem solving, but through a reading of vernacular architecture, the architecture of so-called "unselfconscious cultures," as an exemplar of adaptive architecture: the copying of form again and again while correcting "misfits" that have emerged through use (Alexander 1964: 46). Alexander sought to *simulate* this adaptive process by suppressing the element of time and focusing instead on this process's *structure*. "The residual patterns of adaptive processes," Alexander argued citing the work of cyberneticians Warren Ross Ashby and Norbert Wiener, "are intrinsically well organized" (Alexander 1964: 196). Misfit hierarchies, *trees* derived through the calculation of misfits between physical form and its functional contexts, became the symbol of a well-ordered form-context ensemble akin to the outcome of a temporally unfolding adaptive process. Messy graphs of interacting misfits looking nothing short of overwhelming fell into a neat hierarchy following a mathematical process of graph decomposition.

The tree was as convincing an image of the good organization of a well-adapted system as it was shaky. In the awarded self-corrective article "A City is Not a Tree," Alexander replaced the tree with another topological construct that gave an alternative *image* of well-ordered form-context ensembles: the "semi-lattice" (Alexander 1965). In a hierarchical structure like the tree, the semi-lattice presents the various subsystems as overlapping as opposed to independent. Even though Alexander argued for the semi-lattice with flowery theoretical syllogisms, the semi-lattice's origin was technical. During a consultancy at the MIT Civil Engineering Systems Laboratory, Alexander had developed, in collaboration with Marvin L. Manheim, a computer system called HIerarchical DEcomposition System 2 or HIDECS 2 for ordering graphs of misfits (what Alexander first called "failures") into trees. However, the decomposition of graphs into trees in the system caused "irritating anomalies" (Alexander 1963). Alexander developed a new version of the computer system,

HIDECS 3, that corrected these problems by decomposing graphs into overlapping as opposed to independent subgraphs: by replacing trees with semi-lattices.

Upon becoming a professor at the University of California Berkeley in 1963, Alexander initiated a colossal undertaking to develop conceptual components for the entire modern city (Alexander n.d.). Researchers would collect requirements for different physical parts of the city, analyze their interactions, and develop diagrams. The first project of this ambitious research program was the conceptual design for Rapid Transit Stations in San Francisco. From this project grew the theoretical formulation of what Alexander called "relation": an interaction of physical geometric form with a behavioral, social, or environmental force. Alexander and his collaborators conceptualized relations as interacting and forming the so-called "relational complexes" (Alexander et al. 1966). During a visiting appointment at the UK Ministry of Public Building and Works from 1965 to 1966, Alexander fleshed out these ideas in what he called "relational theory." In it, he defined "relations" as geometric arrangements that prohibited conflicts between people's "tendencies"—what people naturally did when they were given a chance (Alexander and Poyner 1966). These relations were the atoms of what Alexander, in a report co-authored with Barry Poyner, called "environmental structure." Upon his return to the United States in 1967, Alexander co-founded, along with UC Berkeley graduates Sara Ishikawa and Murray Silverstein, a center for the study of environmental structure. The goal of the eponymous Center was to document relations, which were renamed "patterns," and to study their structure in order to produce the so-called "pattern languages."

The first years of the Center for Environmental Structure were dedicated to developing pattern languages for specific projects through small grants and commissions. From this "applied" work emerged a new topological representation of patterns which Center members called the "cascade" (Alexander et al. 1968). The cascade was a network of patterns arranged in a scalar order. In 1968, a large grant from the National Institute of Mental Health allowed the Center to address more "basic," mathematical questions about the structure of the language and its "generativity" (Grabow 1983: 94). Generativity was about the language being able to generate architectural form by relying only on patterns and their rules and not on any external resources that designers brought to bear. *A Pattern Language* was published in 1977 by Oxford University Press and presented 253 patterns, arranged in "straight linear sequence" by order of scale—from regions and territories all the way down to house details (Alexander et al. 1977). The structure of the language was presented as a network—a dense and intricate system that could accommodate multiple different pathways (sequences) for combining the patterns (Alexander 1979: 345) (Fig. 2.3).

In the transition from the "tree" to the mesh-like "network," graphs of different kinds were markers of both continuity and change in theorizing an adaptive relationship between physical form and manifold contexts. Underlying all these shifts in conceptualizing an adaptive architecture—an architecture of good fit—and of the processes for designing it was Alexander's early belief that temporal adaptations are *driven* by invariant structures. This structure encompassed all past actions

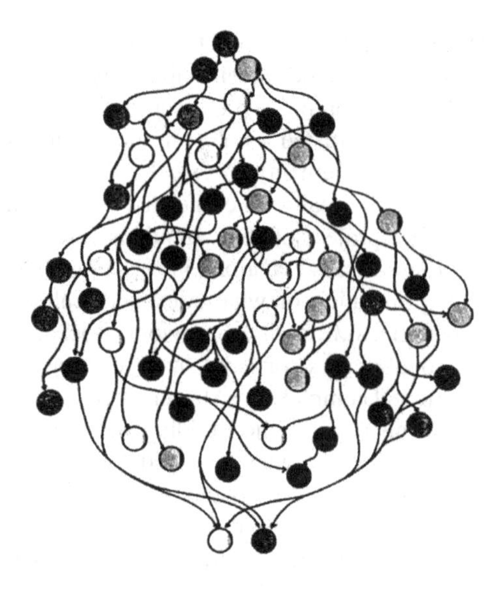

Fig. 2.3 A cascade of patterns at the cover of *A Pattern Language Which Generates Multi-Service Centers*. *Source* Alexander et al. (1968). Reproduced with permission of the Christopher Alexander/Center for Environmental Structure Archives

and circumscribed a plateau of future human agency within the bounds of an all-encompassing organization that remained stable and unchanging despite modifications in individual patterns. The theoretical companion to *A Pattern Language*, titled *The Timeless Way of Building*, is perhaps an unwitting disclosure of this radical suppression of time by understanding an adaptive process as a structural problem: reducing the diachrony of adaptations into the synchrony of trees, semi-lattices, cascades, and networks; into the synchrony of topologies.

2.4 Evolutionary Fit—Entailment Meshes

In 1974, Nicholas Negroponte, founder of the MIT Department of Architecture computing facility that went by the name "the Architecture Machine," gave a presentation at the interdisciplinary symposium *Basic Questions of Design Theory* organized at Columbia University. Entitled "Limits to the Embodiment of Basic Design Theories," the paper chronicled the trajectory of the Architecture Machine from the development of computer-aided design systems sensitive to the designer's idiosyncrasies to programs that enabled non-architects to design their own houses, to what Negroponte referred to as "Intelligent Environments"—a project that the Architecture Machine had initiated in 1969 with a grant from the Graham Foundation for Advanced Study in the Fine Arts. The project was aware of, and responding to, the late-1960s vogue of adaptive architecture that acquired many labels ("flexible," "manipulative," and "responsive") and an unwieldy range of expressions "from the cafetorium to the teepee" (Negroponte 1974: 62).

To this rhetorical and formal medley, Negroponte juxtaposed the notion of "intelligent environments." Unlike regulatory control systems or environments with complex instrumentation like space capsules and cockpits, intelligent environments were characterized by what British cybernetician Gordon Pask had termed the "you-sensor": a combination of predictive models and context-sensitive inferences that emerged from a temporal evolutionary interaction (Negroponte 1974: 64). Predictive models established causal relationships between variables (Lowry 1965: 159) and were often formulated as conditional if–then statements. Although predictive modeling was a standard category in postwar classifications of mathematical and computational models (Lowry 1965; Echenique 1972), the addition of evolutionary learning and statistical inferences made predictive models contingent on the accumulation of data on interactions between occupants and an environment.

A key conceptual influence of the intelligent environments project was Warren Brodey, who in his 1967 article "Intelligent Environments: Soft Architecture" had framed the problem of "evolutionary dialogue" in architectural terms. "How do you design a house," Brodey asked, "which will grow to meet the changes in the family that the house itself will produce?" (1967: 9). The answer was a radical inversion, an involution, which Brodey would return to in the *Radical Software* piece. "Consider the surrounding as the object and man as the environment," Brodey suggested, "or at least make them both object and environment to each other" (1967: 9). Brodey envisioned humans and their environments engaged in rapid cycles of mutual adaptation: "Evolution," he wrote in "Soft Architecture," "now must include evolving environments which evolve man, so that he in turn can evolve more propitious environments in an ever quickening cycle" (1967: 9).

From a computational standpoint, intelligent environments were fertile ground for boosting various developments in the nascent field of Artificial Intelligence, such as facial recognition, natural language processing, and probabilistic inference, to name a few. Their fundamental limit was what Negroponte referred to as their "embodiment." Negroponte characterized Brodey's images as "hackneyed" and "too literal," "brutally transposing ['soft'] from a computational paradigm to a building technology" (1974: 67; 68). "Not everyone wants to live in a balloon," he aphorized (1974: 70). Negroponte saw promise in an unexplored domain of pneumatics that he referred to as "cellular structures." The benefit of these material systems was that their physical structure was amenable to computational transposition. Their form, as Negroponte put, *was* memory (1974: 71). Each cell of these structures would be able to "remember" its past states allowing the structure to develop a model of its ongoing interactions with its occupants. "This can be extrapolated to exercises of cellular automata, in three dimensions," Negroponte speculated, "having the structure dance about" (1974: 71). We can imagine topological representations of the structure as its cells fluidly moved between their "on" and "off" states.

In addition to ideas of dynamic material response that recall contemporary imagery and aspirations of adaptive architecture, the casting of form as memory allowed for predictive and probabilistic calculations of fit based on past interactions between the structure and its occupants thus enabling a kind of *predictive fitness*—a mutual adaptation extrapolated from the record of past interactions. Despite awareness and discussion of adaptable material systems as a difficult problem in achieving intelligent environments, the intelligent environment project's thrust was modeling a fluid interaction between a human occupant and a (materially embedded) computational system. Negroponte captured the requirements of this interaction in his oft-used adage borrowed from Gordon Pask: "My house needs a model of me, a mode of my model of it, and a model of my model of its model of me" (1974: 72). "We know less about how to do this for a house," he added, "than we do for a sketch recognizing machine" (1974: 72).

In 1971, the Architecture Machine had developed a sketch recognition system named HUNCH that syntactically processed sketches hand-drawn on a drawing station (Negroponte 1975: 65). Opposite to other computer drafting systems, such as Ivan Sutherland's SKETCHPAD that distilled the structures and geometric forms behind the user's light pen input, "HUNCH" stored "a voluminous history" of its user's tracings (1975: 65). "The wobbliness of lines, the collections of over tracings, and the darkness of inscriptions," Negroponte wrote, formed the basis on which the system made inferences about the sketch's geometric structure, the designer's intentions, and even preferences in architectural style (1975: 65). SQUINT—an offspring of HUNCH—could additionally recognize boundaries and interpenetrations of shapes thus eliciting "positional" and "proximity" relationships implicit in a hand-drawn sketch of a house plan. These sketch recognition systems were used in combination with space allocation techniques from computer-aided design to devise systems that could circumvent professional architects (Weinzapfel & Negroponte 1976).

In 1976, the Architecture Machine applied for a large grant in collaboration with Richard Bolt of Bolt, Beranek, and Newman (BBN) titled *Computer Mediated Inter- and Intra- Personal Communication.* In the grant proposal, they presented what they called a "graphical conversation theory"—a computer graphics framework based on evolutionary learning between the user and the computational system. The project owed much of its conceptual armature to Gordon Pask, a familiar suspect in art, architectural, and design cycles through publications on cybernetics and design (Pask 1963; 1969) and his participation in the famous UK exhibition *Cybernetic Serendipity* curated by Jasia Reichardt. In his publications on "Conversation Theory," Pask had drawn yet and again fluid blob-like diagrams of conversational interaction between two entities. Among the most evocative of these diagrams was the so-called "entailment mesh," which was featured in graphical conversation theory as a corrective to semantic nets—networks of hard-coded concepts that structured the interactions between a user and a computer system. Pask drew the entailment mesh as a dense, chaotic network of lines between two conversing entities that *emerged* from the conversation: a connective tissue that stitched the two entities together into a single conversational ensemble (Fig. 2.4).

The Architecture Machine's work then presented a telling tension. It combined computer-aided architectural design techniques such as the topological modeling of drawn geometries and the use of graphs to compute permutations of floor plans, with a cybernetic vision of the involution of humans and computational systems.

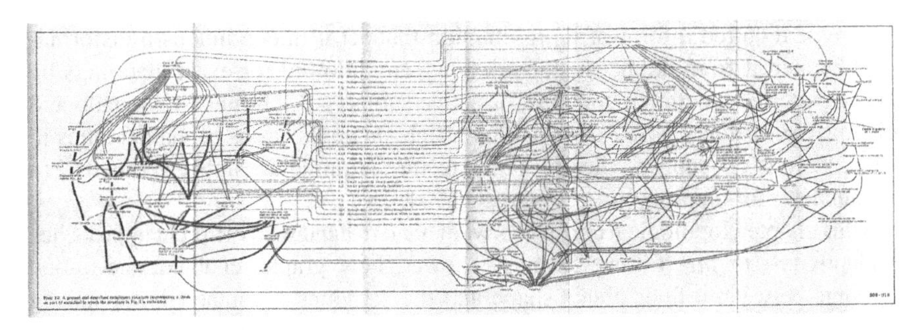

Fig. 2.4 A pruned and described entailment structure (representing a thesis on part of statistics). *Source* Pask (1975)

The virtual blocks of the dialogic computer-aided design system URBAN 5 (Negroponte 1967), the painfully material blocks that (not so successfully) adapted to the behavior of a gerbil colony in Jack Burnham's *Software* show at the New York Jewish Museum (1971: 23), and the geometries of the participatory design research project Architecture-By-Yourself (Weinzapfel & Negroponte 1976) were moved around by graphs that represented functions and behaviors in ways similar to space allocation researchers. But the Architecture Machine also imagined an endless involution between humans and computational systems, stitched together with entailment meshes to achieve Kleinian topologies. In other words, to use Brodey's generative terms, the Architecture Machine envisioned the fluidity of "bioptemes" with the technical repertoire of "mechy max" systems.

2.5 Conclusion

In his 2017 provocation "Everything is Already an Image," John May declared the advent of a "post-orthographic" era in which drawing has been replaced by electronic signal processing (May 2017). Not unlike May, several scholars over the past decades have grappled with the transition from drawing—architecture's erstwhile privileged realm—to the production and manipulation of digital representations (Hewitt 1985; Evans 1997). If classic essays like Robert Bruegmann's "The Pencil and the Electronic Sketchboard" focused on the "aesthetic biases and cultural predispositions" of computer graphics (1989: 151), others have seen computer-propelled changes in architectural representation as unleashing new modes of architectural authorship that rely not on objects but on topological "objectiles" (Carpo 2011: 93). Computers, in these accounts, help enact architectural practices and discourses that orbit around topological abstractions as opposed to concrete geometric particulars. Adaptive architecture is one of them.

This chapter's three episodes collapse histories of adaptive architecture and histories of topological imagination in architecture. For architecture to be imagined as adaptive, it required a form of representation that set it in continuous transformation, liberated it from its metric constrictions, and allowed for exact mappings by activity patterns, functional contexts, or active subjects. As Tomás Maldonado and Gui Bonsiepe astutely observed in a 1964 article on applications of mathematics in design, visions of "rubber" (flexible) architecture required a "rubber geometry" (a colloquialism for topology) to gain thrust (1964: 15).

In this chapter, topological constructs went hand in hand with different approaches to adaptivity—to *fit*. "Fitting Shapes" showed how graphs enabled one-to-one mappings between a floor plan's shape and the activities it contained, supporting endeavors of mutual optimization and giving rise to an enumerative imagination of choice and possibility in design. "Misfit Hierarchies" brought forward an alternative understanding of fit as the elimination of failures, and the use of the relationships between such failures in giving structure to form-context "ensembles." It also highlighted the visual symbolism of topological constructs—from the hierarchical trees to the interconnected networks—as evoking divergent conceptions of environmental complexity and of design processes. Finally, "Evolutionary Fit" presented speculations on "intelligent environments" that, unlike the static patterns or invariant structures of the two former sections, grappled with time and dynamic change. Fit, in this last section, is an equilibrium reached after multiple cycles of mutual adaptation between humans and physical embodiments of computational systems.

The laboratories, architectural researchers, and techniques that I have discussed in this chapter often appear as precedents or inspiration for contemporary research on adaptive architecture. Suffice for the reader to leaf through the contributions of this book to trace influences of ideas from *A Pattern Language* to human–robot interaction (Kahn et al. 2008) and "collaborative environments" (Wang and Green 2019) or more broadly, to appreciate the staying legacy of debates around "intelligent" environmental response as we saw articulated by Negroponte for the field of chitectural robotics (Oosterhuis and Bier 2013) and cyber-physical architecture (Pillan et al. 2020; Lee and Bier 2019; Bier 2016). In fact, much of the early work I have discussed in this chapter figures as a persistent, yet unrealized, vision of how computation, materials, and humans could become part of the same "system"—a relational ensemble of dynamic change through information exchange.

Architectural critics have used terms such as "revival" (Busbea 2020: xvi) or "echo" (Wigley 2001: 114) to refer to such witting (or at times unwitting) adoption of postwar theories and techniques in architects' contemporary engagements with digital technologies. Inarguably, the vast roster of contemporary material implementations and their empirical testing offers valuable insights and new knowledge that move beyond echoes and revivals. However, it seems pressing to activate historical precedents and inspirations of adaptive architecture not as frontiers to conquer but as opportunities for critique—confrontations with the past that help more clearly discern the conceptual and technical apparatus of the broad category of adaptive architecture as a historical, and therefore contingent, construct. Concepts such as "activity," "interaction," "pattern," "conversation," and "evolutionary" that appear in many of the current book's exciting experiments on architectural adaptability can be reframed in their historical specificity, as outcomes of architects' avid engagement with topological representations of form, space, and human inhabitation.

In the three historical episodes presented in this chapter, we see the evocative and instrumental roles of topological constructs in facilitating imaginations of adaptive processes: visualizing the hidden commensurability between activity structures and spatial structures, symbolizing the orders undergirding architectures of "good fit," and stitching together human and machanic agencies in processes of fluid exchange and change. Taken together, these episodes call attention to a core ambivalence in adaptive architecture, concerned as much with architecture's contingencies as it is with new strata of calculative control achieved through new mathematical and computational techniques. If we think of adaptive architecture as an *effect* of architecture's entanglement with topological concepts and images, as opposed to topologies as instruments for fulfilling architectural agendas, we will begin discerning with more clarity implicit theoretical commitments and technical tropes that continue to animate adaptive architecture research in their cultural and political entanglements. In Brodey speak, we might become more aware of the "mechy maxes" technically enacting visions of "bioptemes" and perhaps reconsider the manifold disclosures and unseen aliveness of things that appear to stubbornly always remain just about the same.

References

Alexander C (1963) HIDECS 3: four computer programs for the hierarchical decomposition of systems which have an associated linear graph. Civil Engineering Systems Laboratory Publication Report No, Cambridge, Mass, pp R63–R27

Alexander C (1964) Notes on the synthesis of form. Harvard University Press, Cambridge, Mass

Alexander C (1965) A city is not a tree. Arch. Forum 122:58–62

Alexander C (1967) The question of computers in design. Landscape 14(3):6–8

Alexander C (1979) The timeless way of building. Oxford University Press, New York, N.Y.

Alexander C, King VM, Ishikawa S, Baker M, Hyslop P (1966) Relational complexes in architecture. Archit Rec 140:185–190

Alexander C, Ishikawa S, Silverstein M (1977) A Pattern language: towns, buildings, construction. Oxford University Press

Alexander C, Manheim ML (1962) HIDECS 2: a computer program for the hierarchical decomposition of a set which has an associated linear graph. Cambridge: Civil Engineering Systems Laboratory Publication 160, MIT

Alexander C, Poyner, B (1966) The atoms of environmental structure. Center for Planning and Development Research, University of California, Berkeley, Calif

Alexander C, Ishikawa S, Silverstein M, Center for Environmental Structure (1968) A pattern language which generates multi-service centers. Center for Environmental Structure, Berkeley, Calif

Alexander C (n.d.) "Draft Sent to Serge Chermayeff." Box 32, Folder "Ten Year Program for Research on Environmental Design." Serge Ivan Chermayeff Architectural Records and Papers, 1909–1980. Dept. of Drawings & Archives, Avery Architectural and Fine Arts Library, Columbia University

Banham R (1955) The new brutalism. Arch. Rev. 118:354–361

Bier H (ed) (2016) Special Issue "Cyber-physical Architecture #1, Robotic Building" SPOOL 4 (1)

Bolt RA, Architecture Machine Group (eds) (1976) Computer Mediated Inter- and Intra- Personal Communication. Massachusetts Institute of Technology, Dept. of Architecture, Architecture Machine Group, Cambridge, Mass

Broadbent G (1988) [1973]. Design in architecture: architecture and the human sciences. Fulton, London

Brodey WM (1967) The design of intelligent environments: soft architecture. Landscape 17(1):8–12

Brodey WM (1971) Biotopology 1972. Radic Softw 1(4):4–7

Bruegmann R (1989) The pencil and electronic sketchboard: architectural representation and the computer. In: Blau E, Kaufman E (eds), Architecture and its image: four centuries of architectural representation, works from the collection of the Canadian centre for architecture, Canadian Centre for Architecture, distributed by Cambridge, Mass: The MIT Press

Buffa ES, Armour GS, Vollman TE (1963) Allocating facilities with CRAFT. Harv Bus Rev 42(2):136–140

Busbea L (2007) Topologies: The Urban Utopia in France, 1960–1970. The MIT Press, Cambridge, Mass.; London

Busbea L (2020) The responsive environment: design, aesthetics, and the human in the 1970s. University of Minnesotta Press, Minneapolis, MN

Carpo M (2011) The alphabet and the algorithm. The MIT Press, Cambridge, Mass

Cross N (1977) The automated architect. Pion Ltd., London

Dalrymple Henderson L (2013) The fourth dimension and non-euclidean geometry in modern art, Revised ed. The MIT Press, Cambridge, Mass

Echenique M (1972) Models: A Discussion. In: Martin L, March L (eds) Urban Space and Structures, 164–74. Cambridge Urban and Architectural Studies. Cambridge University Press, New York, N.Y.

Epple M (1998) Topology, Matter, and Space, I: Topological Notions in 19th-century Natural Philosophy. Arch Hist Exact Sci 52:297–392. Springer-Verlag

Evans R (1997) Translations from drawing to building and other essays. The MIT Press, Cambridge, Mass

Friedman Y (1971) The flatwriter: choice by computer. Prog Arch 52:98–101

Friedman Y (1975) Toward a Scientific Architecture. Translated by Cynthia Lang. MIT Press, Cambridge, Mass

Grabow S (1983) Christopher Alexander: the search for a new paradigm in architecture. Routledge Kegan & Paul, Stocksfield ; Boston

Hewitt M (1985) Representational forms and modes of conception; an approach to the history of architectural drawing. J Archit Educ 39(2):2-9. https://doi.org/10.1080/10464883.1985.10758387

Kahn Peter H, Freier Nathan G, Takayuki K, Hiroshi I, Ruckert Jolina H, Severson Rachel L, Kane SK (2008) Design patterns for sociality in human-robot interaction. In: Proceedings of the 3rd ACM/IEEE international conference on Human robot interaction (HRI '08), pp. 97–104. https://doi.org/10.1145/1349822.1349836

Koopmans TC, Tjalling C, Beckmann M (1957) Assignment problems and the location of economic activities. *Econometrica* 25(1):53–76

Kretzer M, Hovestadt L (eds) (2013) ALIVE: Advancements in Adaptive Architecture. Birkhäuser, Basel

Lowry IS (1965) A short course in model design. J Am Inst Plann 31(2):158–166

Maldonado T, Bonsiepe G (1964) "Science and Design." *Ulm* 10/11

March L (2002) Mathematics and Architecture since 1960. In: Rodrigues JF, Williams K (eds) Nexus IV/Centro de Matemática E Aplicaçoes Fundamentais, University of Lisbon, 9–33.

March L, Steadman P (2021) The geometry of environment: an introduction to spatial organization in design. Routledge Revivals

March L, Steadman P (1974) The geometry of environment: an introduction to spatial organization in design. The MIT Press, Cambridge, Mass.

May J (2017) Everything is already an image. Log 40:9–26

Montgomery R (1970) Pattern language - the contribution of Christopher Alexander's centre for environmental structure to the science of design. Arch Forum 132(1):52–59
Moseley DL (1963) A rational design theory for planning buildings based on the analysis and solution of circulation problems. Architects' J 525–537
Muther R (1961) Systematic layout planning. Industrial Education Institute, Boston, Mass
Negroponte N (1967) URBAN 5, an On-Line Urban Design Partner. Ekistics 289–91
Negroponte N (1974) Limits to the embodiment of basic design theories. In Spillers W (ed) Basic questions of design theory, pp. 61–74. North-Holland Pub. Co. ; American Elsevier, Amsterdam; New York, N.Y.
Negroponte N (1975) Soft architecture machines. The MIT Press, Cambridge, Mass.
Osterhuis K, Bier H (2013) IA #5: Robotics in architecture. Jap Sam Books, Prinsenbeek, Netherland
Pask G (1963) The conception of a shape and the evolution of a design. In: Jones JC, Thornley DG (eds) Conference on Design Methods. Pergamon Press, London
Pask G (1969) The architectural relevance of cybernetics. Archit Des 39:494–496
Pask G (1975) Conversation, cognition and learning: a cybernetic theory and methodology. Elsevier, Amsterdam
Pillan M, Bier H, Keith G, Pavlovic M (eds) (2020) Special Issue "Cyber-physical Architecture #3, Actuated and Performative Architecture: Emerging Forms of Human-Machine Interaction" *SPOOL* 7(3)
Rouillard D (2004) *Superarchitecture - Le futur de l'architecture 1950–1970.* Editions de La Villete
Sang L, Bier H (eds) (2019) Special Issue "Cyber-physical Architecture #2, Apparatisation in & of Architecture" SPOOL 6(1).
Steadman P (1970) The automatic generation of minimum-standard house plans. LUBFS Working Paper 23 (March)
Steadman P (1983) Architectural morphology: an introduction to the geometry of building plans. Pion Ltd., London
Steadman P, Brown F, Rickaby P (1991) Studies in the morphology of the english building stock. Environ Plann B Plann Des 18(1):85–98
Steingart A (2015) Inside: Out. Grey Room 59:44–77
Stone PA (1963) Factory Building, Evaluation and Decision: Better Factories. Institute of Directors, London, UK, 249–260
Tabor P (1969) Pedestrian Circulation in Offices. LUBFS Working Paper 17
Teyssot G (2013) Responsive envelopes: the fabric of climatic islands. *Appareil* [Online: http://journals.openedition.org/appareil/1748], https://doi.org/10.4000/appareil.1748
Theodore D (2013) The fattest possible nurse': architecture, computers, and post-war nursing. In Abreu L, Sally S (eds) Hospital Life: Theory and Practice from the Medieval to the Modern. Peter Lang, Oxford/Bern
Vardouli T, Theodore D (2021) Walking instead of working: space allocation, automatic architecture, and the abstraction of hospital labor. IEEE Annals of the History of Computing 43(2): 6–17
Wang Y, Green KE (2019) A pattern-based, design framework for designing collaborative environments. In: Proceedings of the Thirteenth International Conference on Tangible, Embedded, and Embodied Interaction (TEI '19), 595–604. https://doi.org/10.1145/3294109.3295652
Weinzapfel, Guy, and Nicholas Negroponte (1976) Architecture-by-yourself: an experiment with computer graphics for house design. In Proceedings of the 3rd Annual Conference on Computer Graphics and Interactive Techniques. New York, N.Y.: ACM, 74–78
Whitehead B and Eldars MZ (1964). An approach to the optimum layout of single-storey buildings. Architects' J 1373–1380.
Wigley M (2001) Network fever. Grey Room 4:82–122
Witt A (2022) Formulations: architecture, mathematics, culture. The MIT Press, Cambridge, Mass.

Chapter 3
How Do We Want to Interact with Robotic Environments? User Preferences for Embodied Interactions from Pushbuttons to AI

Yixiao Wang and Keith Evan Green

3.1 Introduction

The frontiers of human–computer and human–robot interaction (HCI and HRI) are extending to architecture space, reconfiguring our everyday environments for various activities (Fender and Müller 2019). Robotics is emerging as integral to spatial interactions; however, robot developments for use in the everyday spaces we live in—home, hospital, school, and office—have often focused more on humanoid robotics as replacements for human servants (e.g., Cory and Breazeal 2008) rather than supporting and augmenting human capabilities by forming a collaborative environment (Wang et al. 2019). Nevertheless, in recent years, there has been increasing interest in non-humanoid robotics manifested as robotics-embedded furniture and building systems within everyday spaces (Brauner et al. 2017; Green 2016; Gross and Green 2012; Hoffman et al. 2015; Hoffman and Ju 2014; Ju and Leila 2009; Schafer et al. 2014; Sirkin et al. 2015; Spadafora et al. 2016; Verma et al. 2018). Such robotic artifacts combined with the interior spaces they cohabit are intended to create cyber-physical environments that assist users in daily activities (Sirkin et al. 2015; Verma et al. 2018), augmenting the capacity of users to perform tasks (Schafer et al. 2014; Verma et al. 2018) and even provide them with a semblance of emotional and social support through carefully designed human–robot choreographies (Hoffman et al. 2015; Ju and Leila 2009; Schafer et al. 2014; Sirkin et al. 2015; Verma et al.

Y. Wang (✉)
Design and Artificial Intelligence (DAI), Singapore University of Technology and Design (SUTD), Singapore, Singapore
e-mail: yixiao_wang@sutd.edu.sg

K. E. Green
Human Centered Design (HCD), Sibley School of Mechanical and Aerospace Engineering, Cornell University, Ithaca, NY, USA
e-mail: keg95@cornell.edu

P. Morel and H. Bier (eds.), *Disruptive Technologies: The Convergence of New Paradigms in Architecture*, Springer Series in Adaptive Environments,
https://doi.org/10.1007/978-3-031-14160-7_3

2018). In addition, some design researchers developing non-humanoid robotics have been developing "robot surfaces" as tangible, shape-changing interfaces mediating human–computer interactions (Bosscher and Ebert-Uphoff 2003; Nakagaki et al. 2019; Rosen et al. 2003; Stanley et al. 2016). But these robot surfaces are rarely developed at a larger "environmental" scale or designed as space-making devices (e.g., robotic partitions, ceilings, floors, etc.).

Aligned with this expanded vision (Oosterhuis and Bier 2013) are novel, space-making, robot surfaces (Sirohi et al. 2019; Wang et al. 2019; Wang and Green 2019) which are characterized as malleable, adaptive, physical surfaces that reconfigure interior spaces, supporting and augmenting human activity (Wang et al. 2019; Sturdee and Alexander 2018). These robot surfaces can be embedded in or mounted on ceilings or walls, or be free-standing, and can reconfigure one spatial volume into "many spaces" matched to human activities. The authors envision such robot surfaces having application to confined spaces such as micro-apartments and micro-offices (in costly real estate markets and/or where land is limited), to disaster relief shelters or scientific outposts, to spacecraft and space habitation, and to fully autonomous vehicles (Sirohi et al. 2019).

To provide a sense of robot surface behaviors, we present (in Table 3.1) five "scenarios" for the use-case of designers working in a micro-office. Such compact, physically confined spaces are found increasingly in costly real estate markets and the densest cities due, especially, to both global mass urbanization and the scarcity of land for development. These five scenarios characterize common work activities of the design professions based on former observational studies and literature reviews of designers at work (Wang and Green 2019). Our research team focused on design activities for the scenarios given their mix of digital and manual tool usage and collaborative nature. Design activities encompass wide-ranging kinds of office work, so studying these interactions arguably generalizes to many kinds of collaborative work environments. In preparing these five scenarios, the authors tested with users the question, *What kind of human–robot surface interactions would users prefer most in these scenarios, and why?* For screen-based and other relatively structured tasks, researchers have offered general design guidelines for AI-embedded interface design (Amershi et al. 2019; Höök 2000; Horvitz 1999; Jameson 2008; Norman 1994); but in the wild frontier of spatial human surface interaction, such questions demand considerable attention.

To address the research question, the authors conducted a user study for the micro-office use-case using a robot surface prototype of our design. Conducted in our lab with 12 design major students, our study focused on user experiences with this tangible, interactive system. Participants walked through ("user enactment" by Odom et al. 2012) five scenarios (as per Table 3.1) and selected their preferred interaction modes or proposed new interaction modes as they saw appropriate at key instances in the unfolding activity. Through qualitative analysis, we found significant interaction-mode preference differences for different scenarios and probed the reasons beneath these differences.

Table 3.1 Five "Scenarios" unfolding over the course of common design tasks

Scenario	Conceptual diagram	Task description by scenario
Scenario 1: Notetaking	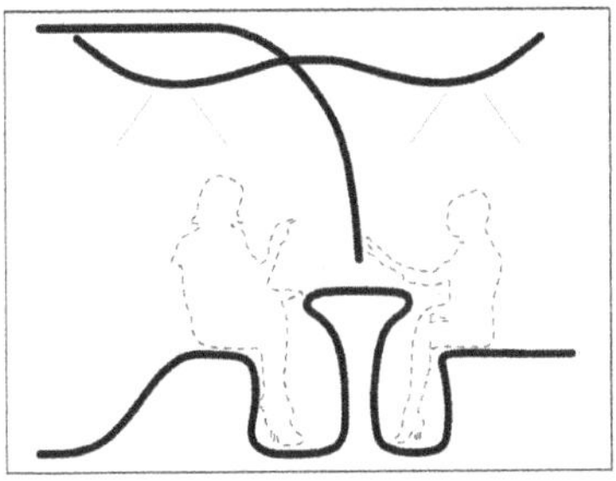	When designers receive comments from clients, reviewers, and fellow designers, they usually take notes for future reference and discussion. Robot surface could help to provide a writing surface (such as a tablet) for notetaking tasks
Scenario 2: Shape and atmosphere simulation	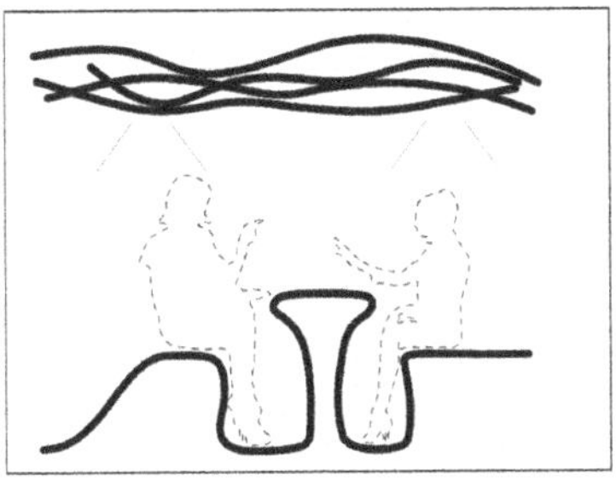	Designers constantly need sources of inspiration during the ideation process. Such inspiration can be an image, a piece of music, a video, a narrative, a conversation, etc. In these scenarios, robot surfaces could use the embedded multimedia systems (e.g., sound, light, and physical movement) to change the atmosphere of the working space. The multimedia environment could serve as a source of design inspiration
Scenario 3: Space division	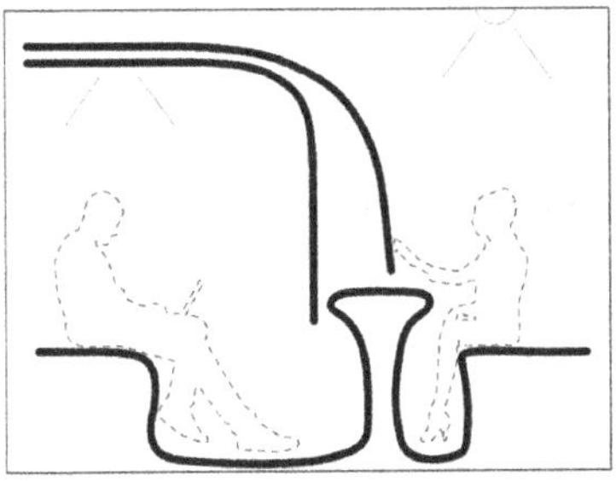	Meetings with clients or reviewers could sometimes be interrupted by something urgent (e.g., urgent email, urgent phone call, family emergency, etc.). In these cases, designers need privacy to handle these situations. Moreover, design activities always consist of a team work and individual work. In either case, robot surfaces could divide the space to create privacy for private workings

(continued)

Table 3.1 (continued)

Scenario	Conceptual diagram	Task description by scenario
Scenario 4: Presentation		Designers give presentations constantly during design iterations, sometimes to clients and sometimes to reviewers. Robot surfaces could help to create a presentation space by temporarily providing big screens, control platforms, and right lighting environment
Scenario 5: Body support		Designers work very hard especially when deadlines are approaching. Many designers work, eat, and even sleep in the design studio. Thus, a comfortable and ergonomic body support, which can be provided by the robot surface, is always welcome and helpful in design studios

3.2 Related Work

Our research of human–robot surface interaction is informed by four topics explored in the literature: *Computer-Supported Cooperative Work* speaking to the environment and context where the robot surface is applied; *Robots that Work with Humans* speaking to our interaction design for robot surfaces and users; *Architectural Robotics* speaking to the capacity of robot surfaces to form physical space serving human needs and wants; and *Shape-changing Interfaces* speaking to the robot surface as interface.

Computer-Supported Cooperative Work

CSCW communities have been exploring computer-supported cooperation for collaborators in different locations through groupware (Lee and Paine 2015; Domova et al. 2013; Campos et al. 2013; Sellen 1992; Tanner and Shah 2010), mixed reality (Billinghurst et al. 2001; Hanaki et al. 2002), and virtual reality (Gauglitz et al. 2014; Greenhalgh and Benford 1995). Our work could arguably be characterized as an exploration of "computer-supported cooperative work environment" in that collaborators in our scenarios (Table 3.1) are working together in a computer-supported office with robot surfaces. The CSCW literature provided us with many insights into the inter-human cooperative working process such as "face-to-face gestural interactions" (Bekker, et al. 1995) and "workspace informal communications" (Sauppé and Mutlu 2014; Whittaker et al. 1994). These insights were useful when we designed interactions for our robot surfaces (Table 3.1). Our work is novel for CSCW research in that the co-workers in our scenario are cooperating in the same physical space, and the work environment reconfiguration occurs physically (i.e., moves physical mass, and not only bits) to support user activities.

Robots that Work with Humans

This research benefits from the well-established body of literature on human–robot interaction (Goodrich 2008), social robotics (Bosscher and Ebert-Uphoff 2003; Fong et al. 2003; Severinson-Eklundh et al. 2003; Tapus et al. 2007), and the ubiquity of robots (Chibani et al. 2013, 2012; Kim et al. 2007), drawing inspiration, especially from the research literature (Schafer et al. 2014; Sirohi et al. 2019; Wang and Green 2019) focused on applications of robots in homes (Forlizzi and DiSalvo 2006; Kidd and Breazeal 2008; Prassler and Kosuge 2008) and healthcare (Kazanzides et al. 2008). Furthermore, our research draws inspiration from research in robotics focused on applications that influence how human beings approach tasks in work environments (Caleb-Solly et al. 2014; Fasola and Mataric 2012; Zawieska and Duffy 2015). Here, the explicit goal is for the robot to incite the user into a different state of activity or consciousness than would not be achieved if the robot were not present. We report here on how robot surface interactions might be made functionally effective, socially supportive, and emotionally encouraging to users in a confined workspace.

Architectural Robotics

We characterize robots that form and reconfigure spaces as "architectural robotics," an emerging subfield in robotics and architectural design (Green 2016). Architectural robotics is, in part, inspired by Malcolm McCullough's vision of "a tangible information commons" in which a "richer, more enjoyable, more empowering, more ubiquitous media become much more difficult to separate from spatial experience" (McCullough 2013; McCullough 2004). Architectural robotics follows, moreover, from the concept of Christopher Alexander et al. of a "compressed pattern" room as elaborated in A Pattern Language (Alexander et al. 1977), conceived for the built environment but since applied to cyber-human systems (Gamma et al. 1995) and human–robot interaction (Kahn et al. 2008). In a compressed pattern room, all the functional rooms within a typical building are essentially formed within a single volume (i.e., one room) that can physically reconfigure.

Following Alexander's compressed pattern room, we envision a robot surface physically reconfiguring (with embedded lighting, audio, sensors, and touch surfaces) to arrive at shape-shifting, functional states supporting and augmenting human activities. Drawing from the scenarios (as per Table 3.1), our robot surfaces configure, in one room, a meeting place, a private working space, a presentation room, and a place of repose. Architectural robotics speaks to the envisioned capacity of a robotic, physical environment to shape wide-ranging human activities.

Shape-Changing Interfaces

Robot surface (Ortega and Goguey 2019) and shape-changing interface (Sturdee and Alexander 2018) are both cyber-physical interfaces that can reconfigure physically. Many shape-changing interfaces, however, are designed specifically for communicating information to users (e.g., physical information displays) and offering dynamic affordances (e.g., shape-changing buttons) (Rasmussen et al. 2012), whereas robot surfaces can be designed to reconfigure the spatial envelops and redefines the spatial affordances for human activities. According to the literature reviews of shape-changing interfaces, architectural applications (Rasmussen et al. 2012; Sturdee et al. 2015), user experiences (Rasmussen et al. 2012, 2013), and user perceptions (Rasmussen et al. 2012), are going to be the future research direction of shape-changing interface. Thus, the investigation into users' preference for space-making robot surfaces reported here is at the frontier of shape-changing surface research.

3.3 Continuum Robot Surface Prototype

As shown in Fig. 3.1, the continuum robot surface is a 2-inch-thick foam panel sectioned by six thin, plywood collars with three tendons running through 3D-printed placeholders mounted to the collars. Three motors mounted at the top of the surface (see Fig. 3.1) pull the tendons to reconfigure the surface in five different configurations that are established as a basic behavioral taxonomy: "rest position," "soft bend," "strong bend," "angled," and "twisted" as shown in Fig. 3.2. Details of the experimentation with the surface configuration and control are reported in Sirohi et al. 2019. This project explores compliant robot surfaces featuring remote actuation of tendons embedded within the surface structure for the following three reasons:

First, as characteristic of continuum robots, tendon-driven continuum robots feature smooth, compliant, and continuously bending bodies inherently suited to operation in close proximity (including interactive and intimate contact) with humans.

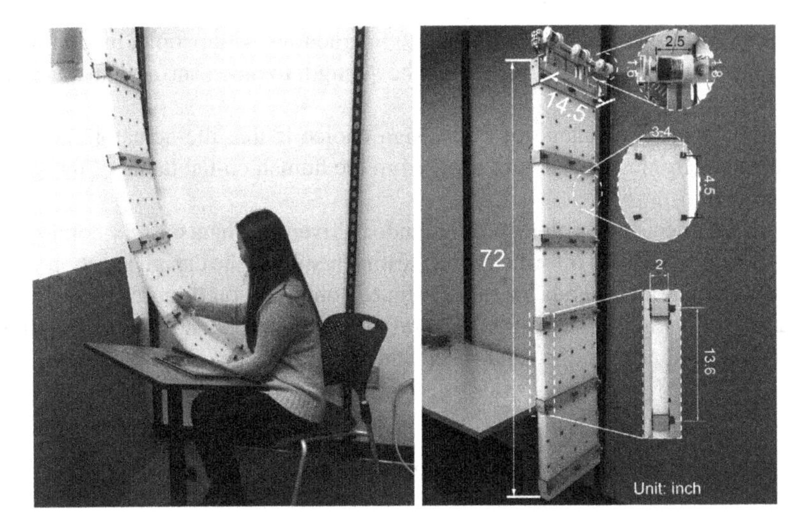

Fig. 3.1 A participant performing task 1 ("Note Taking") with the robot surface prototype in the lab

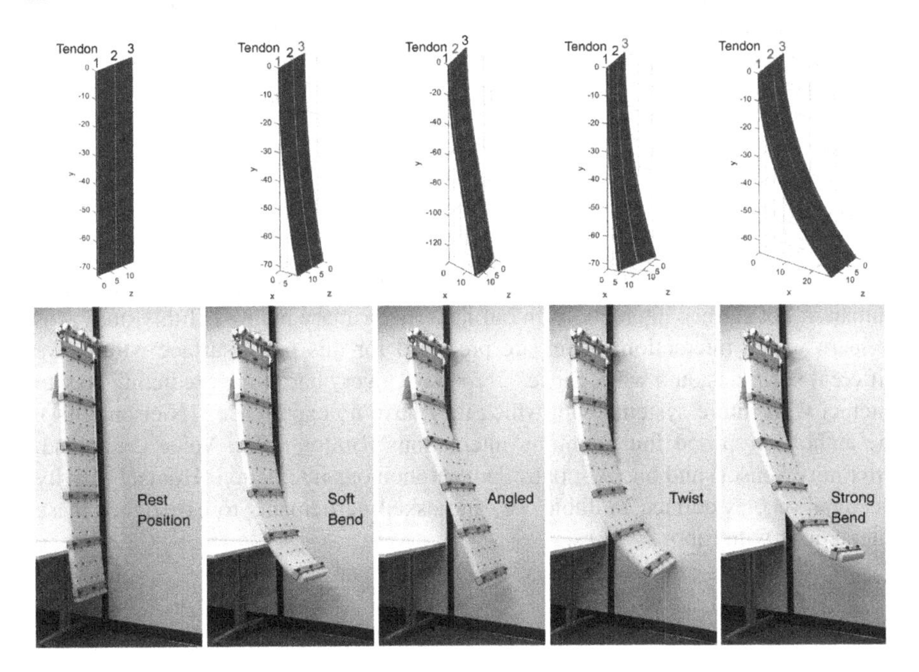

Fig. 3.2 Top row: simulation of 5 configurations, bottom row: prototype images of configurations. Red curves represent tendons being pulled to achieve corresponding configurations

Next, in addition to being well-suited to interactions with people, tendon-driven designs have the advantage of providing the strength to move surfaces that are both large and compliant.

Finally, another advantage of this design choice is that the actuators and their associated electronics can be kept away from the human co-habitants of the shared environment.

The design decision of this specific tendon-driven mechanism (six collars with three brackets on each) is informed by the testing results of the design iteration where tendons were pulled by hands for each designed configuration [Figure 5 from (Sirohi et al. 2019)]. The authors also tried several different motors and 3D-printed motor hubs (Fig. 3.1) to achieve the right speed and strength of the actuating system.

3.4 Physical Prototype In-Lab Study

For this study, we invited to our lab, through convenience sample, 12 university undergraduates and graduates with a design major (interior design, fashion design, and UX design; 5 undergraduates, 7 graduates; ages 18–32; 4 FM, 8 M). Participants provided feedback on how they would like to interact with the robot surface when performing different design tasks and the reasons for their preferences.

Study Design

For this exploratory, qualitative study, the 12 designers were asked to evaluate human–robot surface interactions in our lab. The primary interest of this study was to learn which interaction modes are preferred for this robot surface within five different scenarios, and why? Since "users have a very hard time predicting how to interact with future systems with which they have no experience" (Nielsen 1994), the authors proposed four common interactions (Button, GUI, Voice Command, Proximity Sensor) and one AI-controlled, autonomous interaction (Human Activity Recognition), as defined in Table 3.2, and asked participants to experience these interactions with robot surfaces.

User Enactment and Semi-structured Interview

We conducted user enactments (Odom et al. 2012) by which users "enacted" a scripted scenario, allowing researchers to "observe and probe participants, grounding speculations about how current human values might extend into the future" (Odom et al. 2012). This user experience was followed by semi-structured, in-person interviews with each participant, rewarded with a $10 (USD) Amazon gift card for participating in this 40-min study. One at a time, participants visited our lab fashioned as

Table 3.2 Six interaction modes

Interaction name	Interaction description
Interaction 1: Button (user-controlled)	By pushing the button on the desk, the robot surface will be activated and bend down. By pushing the button again, the robot surface will stop moving
Interaction 2: Voice Command (user-controlled)	By voice commanding Amazon Echo (e.g., "Alexa, provide me a tablet!"), the robot surface will bend down. By voice commanding Amazon Echo (e.g., "Alexa, stop!"), the robot surface will stop where you want it
Interaction 3: Human Activity Recognition (AI-controlled)	In your office, an AI system recognizes your activities using cameras. The AI system is meant to help you with your tasks, anticipating your needs. The AI system is observing your behavioral patterns and tries to understand what you're doing now and what you will do soon. The robot surface will be activated and stopped automatically by AI to assist your work
Interaction 4: Graphic User Interface (user-controlled)	By using a graphic user interface on a touch screen embedded in your worktable, you can control the robot surface. For instance, if you want to bend the robot surface, you can select "strong bend" on the screen
Interaction 5: Proximity Sensor (user-controlled)	There are proximity sensors on the robot surface. You put your hand close to the sensor and it starts to bend. You put your hand close to the sensor again and the robot surface stops where you want it to be
Interaction 6: Anticipatory NLP (AI-controlled, proposed by participants)	In your office, the AI system is listening to your voice and searching for keywords. To anticipate your needs, the AI system is trying to understand what you're doing now and what you will do soon. The robot surface will be activated and stopped automatically by AI to assist your work

a compact, micro-office environment with a chair, table, shelves, and a computer (see Fig. 3.1). In this office setting, we added a functional, button-controlled robot surface prototype of our design measuring (as shown in the same figure). Participants performed work tasks with the prototype as a means to experience human–robot surface interactions as afforded by the surface installed in the work environment. By asking participants to engage in the prescribed work tasks that included the five scenarios (Table 3.1), we allowed the participants to experience both the physical setting of a typical office together with the intervention of our robot surface.

The robot surface configuration used in this user study is initially the "soft bend" (shown in Fig. 3.1) which produces a bend slightly upwards, providing a working surface for writing and reading. In our study, the experimenter controlled the surface using buttons hidden from the participants to simulate different interaction modes for the participants as per the "Wizard of Oz" technique (Dow et al. 2005).

Procedure: Step-by-Step Study Protocol and Interview Questions

1. First, we introduced this robot surface prototype to the participant by verbally describing its structures, functions, and potential applications.
2. Second, we presented a video showing five different configurations that could be assumed by the robot surface prototype, including "Resting Position," "Strong Bend," "Soft Bend," "Twist," and "Angled." This video helped the participant better understand the functionality of the robot surface and the context of the study.
3. Third, we introduced five ways in which people could interact with the robot surface: Buttons, Voice Command, Human Activity Recognition, GUI Interface, and Proximity Sensors, as specified in Table 3.2.
4. Fourth, we introduced five scenarios (Table 3.1) and asked the participant to role-play, initially, scenario 1. For scenario 1 (i.e., notetaking during a conversation with a client), the participant performed the task with the robot surface prototype in our lab (Fig. 3.1). Using the "Wizard of Oz" technique Spadafora et al. (2016), we simulated all five interactions, one by one, for the participant while he/she is performing the given task. We then asked the participant to choose her/his most preferred interactions or propose new interactions.
5. Fifth, after experiencing the five interactions through role-playing scenario 1, the participants were each presented with videos, pictures, and narrative descriptions of scenarios 2, 3, 4, and 5 (Table 3.1). For each scenario, after watching the corresponding videos and pictures with narratives, the participant chose his/her favorite interactions or propose new interactions if none of the 5 interactions is preferred. The participant then offered reasons for his/her preferences.
6. Sixth, our study was followed by a semi-structured interview with three open-ended questions. Notes were taken to record answers offered by the participant. Question 1: *What is your impression of this technology?*; Question 2: *What are things you would improve?*; and Question 3: *What other use cases can you think of for this technology?*.

Qualitative Data Analysis and Results

Each participant was asked to choose their favorite interaction modes for each scenario. Some participants chose one or two interaction modes as their favorite, and others proposed new interaction modes for some scenarios since none of the five interaction modes was preferred. Figure 4 shows how many times each of the five interaction modes was voted as the favorite one for each scenario by participants. The data is color coded with deeper blue representing more votes.

Participants' Feedback on "Human Activity Recognition" (Interaction 3)

As visualized in Fig. 3.3, the most preferred interaction mode overall for the 12 participants is "Human Activity Recognition," which was the most voted interaction for scenarios 1, 4, and 5. Below are participants' feedback on this interaction, expressed as a list of written statements which, for us, captures what was communicated repeatedly by two or more participants.

1. The mental load using "Human Activity Recognition" is lower than other interaction modes; therefore, users are less likely to be interrupted in their tasks (as offered by Participants 1, 2, 3, 11, 12).
2. Human Activity Recognition is "most convenient" for accomplishing simple tasks, as the system's intelligence saves the human user the chore of giving specific commands or instructions to the system (Participants 1, 6, 9, 10).
3. By capturing data coming with human body gestures as a control mode, more comfortable body support could be provided (Participants 5, 8, 11).
4. It is "discreet rather than distracting" to the user (Participants 7, 10).
5. The interaction process feels natural, as if the robot surface is the body extension of oneself (Participants 2, 8).

Additionally, some participants were concerned that interaction by Human Activity Recognition might not be accurate enough (Participants 5, 8, 9), smart enough (Participant 10), or offer users sufficient control of the system (Participant 12).

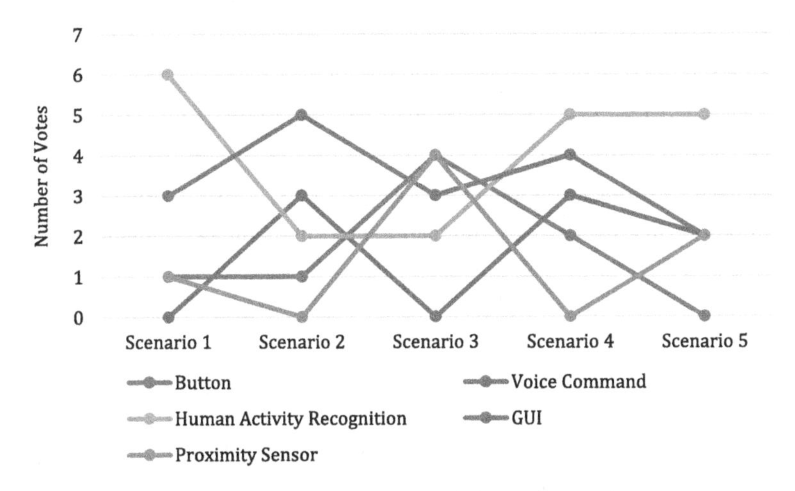

Fig. 3.3 Number of votes on favorite interaction modes for each scenario

Participants' Feedback on "Graphic User Interface" (Interaction 4)

The second most preferred interaction mode for the 12 participants, overall, was "Graphic User Interface," which was also the most voted interaction for scenario 2 and the second most voted interaction for scenarios 1 and 4. Below are participants' feedback on this interaction:

1. This interaction is relatively quiet, discreet, and not distracting (Participants 2, 5, 6, 10).
2. This interaction offers user control with enough many options (Participants 2, 12)
3. This interaction offers a simpler, easier, and more familiar control over the system (Participant 8).

Additionally, participants 1, 4, 8, 10, and 12 suggested that they would like to see graphic sliders instead of only buttons for fine-tuning the robot surface's bending angle and icons to tap as shortcuts to control predefined robot surface configurations. These are useful suggestions for us to consider for further user studies and prototype iterations.

Participants' Feedback on "Voice Command" (Interaction 2)

The third most preferred interaction mode for the 12 participants was "Voice Command," which was also the second most voted interaction in scenarios 2 and 4. Below are participants' feedback on this interaction mode:

1. Voice Command allows the user to give commands with the least effort while multi-tasking (Participants 6, 12).
2. Talking to a robot in front of other people is natural and straightforward (Participants 2, 9, 11).
3. Voice Command allows you to control freely with much more options than other interactions (Participants 2, 4, 12).
4. Language used to convey commands could convey specific meanings to the system (Participants 2, 3).

Additionally, participants 3, 10, 11, and 12 were concerned that talking to the robot surface might be disruptive to accomplishing tasks and human–human interactions. Furthermore, participant 2 mentioned that Voice Command might not be convenient when users communicate with someone else via phone.

Participants' Feedback on "Proximity Sensor" (Interaction 5)

"Proximity Sensor" overall ranked the 4th most preferred interaction mode, and was also one of the two most voted interactions for scenario 3. Participants 1, 5, 6, 7, 10, and 11 suggested that Proximity Sensors are a direct, reliable, and tangible interaction. Participants 7 and 12 suggested that it is natural to use Proximity Sensors for something urgent or time sensitive. Meanwhile, participants 2 and 10 reported that Proximity Sensors could cause interference and distraction given that users need to keep watching the robot surface before touching the sensor a second time to stop it. Finally, Participant 12 suggested that Proximity Sensors offer very few options to control the surfaces.

Participants' Feedback on "Button" (Interaction 1)

"Button" ranked the fifth most preferred interaction mode in total votes and was one of the two most voted interactions for scenario 3. Participants in favor of this interaction suggested that buttons are simple, straightforward, and intuitive (Participants 5 and 12). Participants 1 and 6 also mentioned that buttons were more discreet and less disruptive in human–human interaction, especially in the occasionally awkward social situation that at times occurs at work. Meanwhile, two participants argued that buttons are too cumbersome and not "that beautiful" (Participants 2 and 3).

Participants' Identification of Interaction Modes to Add/Consider

Some participants recommended other interfaces that might suit the five scenarios described. "Anticipatory Natural Language Processing" (Interaction 6, Table 3.2) was a new interaction mode proposed by multiple participants (5 out of 12 participants) mostly for scenario 3. Participants in favor of this interaction suggested that an interaction based on the system picking-up verbal cues instead of requiring direct commands issued by the user makes life easier (Participants 1 and 9). Four participants also argued that it feels natural in situations such as captured in scenario 3 (Participants 3, 4, 9, and 11). The authors believe that this is an important interaction mode that should be carefully considered for future design.

Two participants proposed "Pressure Sensing" for scenario 5, where they argued that the robot surface should provide back support intelligently by adjusting its curvature ergonomically based on the amount of pressure received by the AI system from the sensor grid embedded in the robot surface. One participant proposed, as well, Joystick" for scenario 1, as he/she preferred "a more tangible version of GUI interface." The authors believe these are all inspiring ideas for designing human-surface interaction.

3.5 Findings and Discussion

We now consider how our results (4.1) provide insights for robot surface interaction design in a compact, interactive working space, and (4.2) inspire future research for complex human-surface interactions including those where AI is embedded in the built environment.

Insights for Robot Surface Interaction Design

The interaction modes that will be discussed here include AI-controlled interactions (Human Activity Recognition, Anticipatory NLP, and Pressure Sensing) and user-controlled interactions (Button, GUI, Voice Command, Proximity Sensor, Joystick). Here, AI-controlled interactions refer to the interaction modes where the AI-embedded system automatically gathers information from the users (e.g., from users' working activities, verbal cues, and body postures), analyzes the data, and makes decisions on activating or reshaping the robot surface for users. User-controlled interactions refer to the interactions where users give the direct command to the system:

1. Users prefer AI-controlled modes for the simpler scenarios (e.g., Scenario 5 "body support") which require fewer control options or complexities. For simple scenarios, people would like "the system's intelligence to save the chore of giving specific commands to the system" (as commented by participants 1, 6, 9, 10). On the other hand, Scenario 2 ("Shape and Atmosphere Simulation") is a complex task requiring more control of alternative surface reconfigurations, which is perhaps why users choose "GUI Interface" and "Voice Control" to acquire more control over the system (as commented by participants 2, 3, 4, 12).
2. Users prefer AI-controlled modes for scenarios where they prefer the human-surface interactions to happen in a discreet, or natural way with instant feedback as if the surfaces are extensions of oneself. For instance, in scenarios 1 and 4, users want the tablet or presentation screen to be delivered by the surface without interrupting the conversation (Participants 1, 3, 6, 10, 11); in scenario 3, users want the robot surface to divide the space automatically after they excused themselves for urgent emails or phone calls from the clients (participants 3, 4, 9, 10); in scenario 5, users proposed the "Pressure Sensing" interaction modes so that they can get instant feedback and constantly change the robot surface curvature with a more comfortable body position (participants 1, 3). However, in some scenarios (scenarios 3 and 4), "GUI interface" can also be described as "relatively discreet, quiet, and not distracting" (Participants 2, 5, 6, 10).
3. For the controls that cannot be easily specified by direct commands (such as the detailed curvature of the robot surface), users prefer the system to gather detailed information by itself and then reconfigure the surface properly. For instance, in scenario 5, users prefer the AI system to gather pressure data automatically and

reconfigure the robot surface curvature to fit body postures, since they believe it is easier and more precisely controlled in this way (participants 3, 5, 6, 8). For controls that can be easily specified and described, however, users prefer "Voice Command" when "discreetness" is not part of the equation, since it is natural and straightforward (participants 2, 4, 9, 11, 12).

Nevertheless, there are some concerns with AI-controlled interactions, including the system's control accuracy (participants 5, 8, 9, 11) and its ability to correctly interpret the situation (participants 2, 9, 10, 11). In short, there are trust issues with the AI-controlled system. On the other hand, users are usually more familiar with and confident about user-controlled interaction modes. Designers should carefully take these aspects into consideration when designing spatial human-surface interactions.

Future Research

Because of the limited time, we could devote to each participant session in our lab study and the limited number of interaction modes participants could remember when making a choice, we elected with hesitation to not include semi-autonomous interaction modes as an option in our studies. Interestingly, users didn't propose any semi-autonomous interactions either in the study. We intend to pursue this research direction in the future as we intensively conduct further user studies with a full-functioning system of multiple robot surfaces (Houben et al. 2016). We will explore user preferences for semi-autonomous interfaces with built-in verification steps (Höök 2000; Norman 1994), direct manipulation constructs (Horvitz 1999), and predictable AI behaviors (Amershi et al. 2019). The authors believe a seamless integration of AI-controlled, user-controlled, and semi-autonomous interactions can be the next step of spatial human-surface choreography in an interactive space.

Informed by our findings, we will construct a compact office space with up to three fully functional robot surfaces enabling different interactions for predefined scenarios. The results of the studies reported here will inform the interaction modes we will implement in the next prototype. Additionally, the number of robot surfaces (one, two, three?) will also be a variable intended for our further study. We will invite participants with different backgrounds to perform defined scenarios with, and without the robot surfaces to again characterize human–robot surface interactions (user experience, usability) and also, this next time, compare task performance (efficacy) under treatment and control conditions (e.g., number of errors made by participants, number of examples produced, quality of examples produced as judged by experts).

3.6 Conclusion

In this paper, we identified user preferences for human–robot surface interactions, from pushbuttons to AI, within a compact (small volume) physical space through a qualitative, in-lab study using a robotic surface of our own design. The outcome of our study may be viewed as validation of what Gordon Pask and Nicholas Negroponte suggested decades ago (Pask 1969; Negroponte 1975); that intelligence in architecture emerges through socially intelligent interaction when both sides of the interaction are intelligent.

Our design and the outcomes of our study provide architects, roboticists, and human–computer interaction researchers an understanding of the complex intelligence that emerges through interactions between humans and AI-embedded architectural robotics. Specifically, the outcomes of our user study offer designers knowledge about user preferences for shape-changing surfaces and spaces with different autonomy levels as found in realistic, working-life scenarios. More broadly, this research informs a deeper understanding of our coexistence with robot and AI-embedded built environments. Such environments manifested as physically reconfigurable micro-offices, micro-apartments, and assistive care facilities are likely to proliferate as society continues to mass-urbanize, grow older, and grow in numbers.

References

Alexander C, Ishikawa S, Silverstein M (1977) A pattern language: towns, buildings, construction. Oxford University Press

Amershi S, Weld D, Vorvoreanu M, Fourney A, Nushi B, Collisson P, Suh J, Iqbal S, Bennett PN, Inkpen K, Teevan J, Kikin-Gil R, Horvitz E (2019) Guidelines for human-AI interaction. In: Proceedings of the 2019 CHI conference on human factors in computing systems (CHI '19), Paper 3, 13 p. https://doi.org/10.1145/3290605.3300233

Bekker MM, Olson JS, Olson GM (1995) Analysis of gestures in face-to-face design teams provides guidance for how to use groupware in design. In: Proceedings of the conference on designing interactive systems processes, practices, methods, & techniques (DIS '95)

Billinghurst M, Billinghurst M, Kato H, Poupyrev I (2001) MagicBook: transitioning between reality and virtuality. In: CHI '01 extended abstracts on human factors in computing systems (CHI EA '01), pp 25–26

Bosscher P, Ebert-Uphoff I (2003) Digital clay: architecture designs for shape-generating mechanisms. In: 2003 IEEE international conference on robotics and automation (Cat. No.03CH37422). https://doi.org/10.1109/robot.2003.1241697

Brauner P, van Heek J, Ziefle M, Hamdan NA-H, Borchers J (2017) Interactive FUrniTURE: evaluation of smart interactive textile interfaces for home environments. In: Proceedings of the 2017 ACM international conference on interactive surfaces and spaces (ISS '17). Association for Computing Machinery, New York, NY, USA, pp 151–160. https://doi.org/10.1145/3132272.3134128

Caleb-Solly P, Dogramadzi S, Ellender D, Fear T, van den Heuvel H (2014) A mixed-method approach to evoke creative and holistic thinking about robots in a home environment. In: Proceedings of the 2014 ACM/IEEE international conference on human-robot interaction (HRI '14), pp 374–381. https://doi.org/10.1145/2559636.2559681

Campos P, Ferreira A, Lucero A (2013) Collaboration meets interactive surfaces: walls, tables, tablets, and phones. In Proceedings of the 2013 ACM international conference on Interactive tabletops and surfaces (ITS '13). Association for Computing Machinery, New York, NY, USA, pp 481–482. https://doi.org/10.1145/2512349.2512350

Chibani A, Amirat Y, Mohammed S, Hagita N, Matson ET (2012) Future research challenges and applications of ubiquitous robotics. In: Proceedings of the 2012 ACM conference on ubiquitous computing (UbiComp '12), pp 883–891. https://doi.org/10.1145/2370216.2370415

Chibani A, Amirat Y, Mohammed S, Matson E, Hagita N (2013) Barreto M (2013) ubiquitous robotics: recent challenges and future trends. Robot Auton Syst 61(11):1162–1172

Cory KD, Breazeal C (2008) Robots at home: understanding long-term human-robot interaction. In: Proceedings of international conference on intelligent robots and systems (IEEE/RSJ '08), pp 3230–3235

Domova V, Vartiainen E, Azhar S, Ralph M (2013) An interactive surface solution to support collaborative work onboard ships. In: Proceedings of the 2013 ACM international conference on Interactive tabletops and surfaces (ITS '13). Association for Computing Machinery, New York, NY, USA, pp 265–272. https://doi.org/10.1145/2512349.2512804

Dow S, Lee J, Oezbek C, MacIntyre B, David Bolter J, Gandy M (2005) Wizard of Oz interfaces for mixed reality applications. In: CHI '05 extended abstracts on human factors in computing systems (CHI EA '05), pp 1339–1342. https://doi.org/10.1145/1056808.1056911

Fasola J, Mataric MJ (2012) Using socially assistive human–robot interaction to motivate physical exercise for older adults. In: Proceedings of the IEEE, 100, 8 (August 2012), pp 2512–2526

Fender A, Müller J (2019) SpaceState: Ad-Hoc definition and recognition of hierarchical room states for smart environments. In: Proceedings of the 2019 ACM international conference on interactive surfaces and spaces (ISS '19). Association for Computing Machinery, New York, NY, USA, pp 303–314. https://doi.org/10.1145/3343055.3359715

Fong T, Nourbakhsh I, Dautenhahn K (2003) A survey of socially interactive robots. Robot Auton Syst 42(3-4):143–166

Forlizzi J, DiSalvo C (2006) Service robots in the domestic environment: a study of the roomba vacuum in the home. In: Proceedings of the 1st ACM SIGCHI/SIGART conference on Human-robot interaction (HRI '06), pp 258–265. https://doi.org/10.1145/1121241.1121286

Gamma E, Helm R, Johnson R, Vlissides J, Booch G (1995) Design patterns: elements of reusable object-oriented software. Addison-Wesley Longman Publishing Co. Inc.

Gauglitz S, Nuernberger B, Turk M, Höllerer T (2014) World-stabilized annotations and virtual scene navigation for remote collaboration. In: Proceedings of the 27th annual ACM symposium on User interface software and technology (UIST '14), pp 449–459. https://doi.org/10.1145/2642918.2647372

Goodrich MA, Schultz AC (2008) Human–robot interaction: a survey. Found Trends Hum–Comput Interact 1(3):203–275

Green KE (2016) Architectural robotics: ecosystems of bits, bytes, and biology. MIT Press

Greenhalgh C, Benford S (1995) MASSIVE: a collaborative virtual environment for teleconferencing. ACM Trans Comput-Hum Interact 2(3):239–261. https://doi.org/10.1145/210079.210088

Gross MD, Green KE (2012) Architectural robotics, inevitably. Interactions 19(1):28–33. https://doi.org/10.1145/2065327.2065335

Hanaki S-I, Tansuriyavong S, Endo M (2002). Experiences in VILLA: a mixed reality space to support group activities. In: Proceedings of the 4th international conference on Collaborative virtual environments (CVE '02), pp 155–156. https://doi.org/10.1145/571878.571907

Hoffman G, Ju W (2014) Designing robots with movement in mind. J Hum-Robot Interact 3(1):91–122

Hoffman G, Zuckerman O, Hirschberger G, Luria M, Shani Sherman T (2015) Design and evaluation of a peripheral robotic conversation companion. In: Proceedings of the tenth annual ACM/IEEE international conference on human-robot interaction (HRI '15), pp 3–10. https://doi.org/10.1145/2696454.2696495

Höök K (2000) Steps to take before intelligent user interfaces become real. Interact Comput 12(4):409–426

Horvitz E (1999) Principles of mixed-initiative user interfaces. In: Proceedings of CHI '99. ACM, New York, NY, USA, pp 159–166

Houben S, Vermeulen J, Klokmose C, Schöning J, Marquardt N, Reiterer H (2016) Cross-surface: challenges and opportunities of spatial and proxemic interaction. In: Proceedings of the 2016 ACM international conference on interactive surfaces and spaces (ISS '16). Association for Computing Machinery, New York, NY, USA, pp 509–512. https://doi.org/10.1145/2992154.2996360

Jameson A (2008) Adaptive interfaces and agents. In: Sears A, Jacko JA (eds) The human-computer interaction handbook: fundamentals, evolving technologies and emerging applications (2nd ed) CRC Press, Boca Raton, FL, pp 433–458

Ju W, Leila T (2009) Approachability: how people interpret automatic door movement as gesture. Int J Des 3(2):1–10

Kahn PH, Freier NG, Kanda T, Ishiguro H, Ruckert JH, Severson RL, Kane SK (2008) Design patterns for sociality in human-robot interaction. In: Proceedings of the 3rd ACM/IEEE international conference on Human robot interaction (HRI '08), pp 97–104. https://doi.org/10.1145/1349822.1349836

Kazanzides P, Fichtinger G, Hager GD, Okamura AM, Whitcomb LL, Taylor RH (2008) Surgical and interventional robotics-core concepts, technology, and design. IEEE Robot Automat Mag 15(2):122–130

Kidd CD, Breazeal C (2008) Robots at home: understanding long-term human-robot interaction. In: IEEE/RSJ international conference on intelligent robots and systems (ICRA '08), pp 3230–3235

Kim J-H, Lee K-H, Kim Y-D, Suresh Kuppuswamy N, Jo J (2007) Ubiquitous robot: a new paradigm for integrated services. In: Proceedings of IEEE international conference on robotics and automation (ICRA '07), pp 2853–2858

Lee CP, Paine D (2015) From the matrix to a model of Coordinated Action (MoCA): a conceptual framework of and for CSCW. In: Proceedings of the 18th ACM conference on computer supported cooperative work & social computing (CSCW '15), pp 179–194. https://doi.org/10.1145/2675133.2675161

McCullough M (2004) Digital ground: architecture, pervasive computing, and environmental knowing. MIT press

McCullough M (2013) Ambient commons: Attention in the age of embodied information. MIT Press

Nakagaki K, Fitzgerald D, (John) Ma Z, Vink L, Levine D, Ishii H (2019) inFORCE: Bi-directional 'force' shape display for haptic interaction. In: Proceedings of the thirteenth international conference on tangible, embedded, and embodied interaction (TEI '19). ACM, New York, NY, USA, pp 615–623. https://doi.org/10.1145/3294109.3295621

Negroponte N (1975) Soft architecture machines. MIT Press, Cambridge, Mass./London

Nielsen J (1994) Executive summary. Usability engineering. Morgan Kaufmann, San Diego, CA, pp 11–12

Norman DA (1994) How might people interact with agents. Commun ACM 37(7):68–71

Odom W, Zimmerman J, Davidoff S, Forlizzi J, Dey AK, Kyung Lee M (2012) A fieldwork of the future with user enactments. In: Proceedings of the designing interactive systems conference (DIS '12), pp 338–347. https://doi.org/10.1145/2317956.2318008

Oosterhuis K, Bier H (2013) IA #5: robotics in architecture. Jap Sam Books, Prinsenbeek, Netherland

Ortega M, Goguey A (2019) BEXHI: a mechanical structure for prototyping bendable and expandable handheld interfaces. In: Proceedings of the 2019 ACM international conference on interactive surfaces and spaces (ISS '19). Association for Computing Machinery, New York, NY, USA, pp 269–273. https://doi.org/10.1145/3343055.3359703

Pask G (1969) The architectural relevance of cybernetics. Archit Des 39(9):494–496

Prassler E, Kosuge K (2008) Domestic robotics. In: Siciliano B, Khatib O (eds) Springer handbook of robotics. Springer, pp 1253–1281

Rasmussen MK, Pedersen EW, Petersen MG, Hornbæk K (2012) Shape-changing interfaces: a review of the design space and open research questions. In: Proceedings of the SIGCHI conference on human factors in computing systems (CHI '12). Association for Computing Machinery, New York, NY, USA, pp 735–744. https://doi.org/10.1145/2207676.2207781
Rasmussen KM, Grönvall E, Kinch S, Graves Petersen M (2013) "It's alive, it's magic, it's in love with you": opportunities, challenges and open questions for actuated interfaces. In: Proceedings of the 25th Australian computer-human interaction conference: augmentation, application, innovation, collaboration (OzCHI '13). Association for Computing Machinery, New York, NY, USA, pp 63–72. https://doi.org/10.1145/2541016.2541033
Rosen D, Nguyen A, Wang H (2003) On the geometry of low degree-of-freedom digital clay human-computer interface devices. In: Proceedings of 2003 international design engineering technical conferences and computers and information in engineering conference (ASME '03), pp 1135–1144
Sauppé A, Mutlu B (2014) How social cues shape task coordination and communication. In: Proceedings of the 17th ACM conference on computer supported cooperative work & social computing (CSCW '14). https://doi.org/10.1145/2531602.2531610
Schafer G, Green K, Walker I, King Fullerton S, Lewis E (2014) An interactive, cyber-physical read-aloud environment: results and lessons from an evaluation activity with children and their teachers. In: Proceedings of the 2014 conference on Designing interactive systems (DIS '14), pp 865–874. https://doi.org/10.1145/2598510.2598562
Sellen AJ (1992) Speech patterns in video-mediated conversations. In: Bauersfeld P, Bennett J, Lynch G (eds) Proceedings of the SIGCHI conference on human factors in computing systems (CHI '92), pp 49–59. https://doi.org/10.1145/142750.142756
Severinson-Eklundh K, Green A, Hüttenrauch H (2003) Social and collaborative aspects of interaction with a service robot. Robot Autonom Syst 42(3-4):223–234
Sirkin D, Mok B, Yang S, Ju W (2015) Mechanical ottoman: engaging and taking leave. In: Proceedings of the tenth annual ACM/IEEE international conference on human-robot interaction extended abstracts (HRI'15 Extended Abstracts), pp 275–275. https://doi.org/10.1145/2701973.2702096
Sirohi, R, Wang, Y., Hollenberg, S., Godage, I. S., Walker, I. D., and Green, K. E. 2019. Design and Characterization of a Novel, Continuum-Robot Surface for the Human Environment. In Proceedings of the 15th Conference on Automation Science and Engineering (IEEE CASE 2019), August 22-26, Vancouver, BC, pp. 1169–1174
Spadafora M, Chahuneau V, Martelaro N, Sirkin D, Ju W (2016) Designing the behavior of interactive objects. In: Proceedings of the TEI '16: tenth international conference on tangible, embedded, and embodied interaction (TEI '16), pp 70–77. https://doi.org/10.1145/2839462.2839502
Stanley AA, Hata K, Okamura AM (2016) Closed-loop shape control of a haptic jamming deformable surface. In: 2016 IEEE international conference on robotics and automation (ICRA). https://doi.org/10.1109/icra.2016.7487433
Sturdee M, Alexander J (2018) Analysis and classification of shape-changing interfaces for design and application-based research. ACM Comput Surv 51(1):32. Article 2. https://doi.org/10.1145/3143559
Sturdee M, Hardy J, Dunn N, Alexander J (2015) A public ideation of shape-changing applications. In: Proceedings of the 2015 international conference on interactive tabletops & surfaces (ITS '15). Association for Computing Machinery, New York, NY, USA, pp 219–228. https://doi.org/10.1145/2817721.2817734
Tanner P, Shah V (2010) Improving remote collaboration through side-by-side telepresence. In: CHI '10 extended abstracts on human factors in computing systems (CHI EA '10), pp 3493–3498. https://doi.org/10.1145/1753846.1754007
Tapus A, Maja M, Scassellatti B (2007) The grand challenges in socially assistive robotics. IEEE Robot Autom Mag, Institute of Electrical and Electronics Engineers 14(1)
Verma S, Gonthina P, Hawks Z, Nahar D, Brooks JO, Walker ID, Wang Y, de Aguiar C, Green KE (2018) Design and evaluation of two robotic furnishings partnering with each other and their users to enable independent living. In: Proceedings of the 12th EAI international conference on

pervasive computing technologies for healthcare (PervasiveHealth '18), pp 35–44. https://doi.org/10.1145/3240925.3240978

Wang Y, Frazelle C, Sirohi R, Li L, Walker ID, Green KE (2019) Design and characterization of a novel robotic surface for application to compressed physical environments. In: 2019 international conference on robotics and automation (ICRA '19), pp 102–108

Wang Y, Green KE (2019) A pattern-based, design framework for designing collaborative environments. In: Proceedings of the thirteenth international conference on tangible, embedded, and embodied interaction (TEI '19), pp 595–604. https://doi.org/10.1145/3294109.3295652

Whittaker S, Frohlich D, Daly Jones O (1994) What is it like and how might we support it? In: Conference companion on human factors in computing systems (CHI '94). https://doi.org/10.1145/259963.260328

Zawieska K, Duffy BR (2015) The social construction of creativity in educational robotics. In: Progress in automation, robotics and measuring techniques. Springer International Publishing, pp 329–338

Chapter 4
Design-to-Robotic-Production and -Operation for Activating Bio-Cyber-Physical Environments

Henriette Bier, Arwin Hidding, Max Latour, Pierre Oskam, Hamed Alavi, and Alara Külekci

4.1 Urban Context

Residual spaces resulting from inter al. abandonment of vernacular or industrial buildings due to de-industrialization, migration, political and economic shifts, and ineffective planning (Accordino and Johnson 2000; Haase et al. 2016; Oskam et al. 2021) contain valuable assets, for instance, unique animal and plant species that inhabit such abandoned places (inter al. Laurie 1979). These places offer potential environments for wildlife and natural growth within the urban fabric (Kawata 2014), accommodating species that often find no place because of inter al. intensified agriculture (Harrison and Davies 2002; Kowarik 2013; Schwarz 1980). Furthermore, they serve as meeting places for the youth engaging in artistic creation, play, and exploration (Edensor 2005). Residual spaces introduce thus new opportunities for material and social interaction although their ecosystems remain fragile and they may often be misused for illegal activities. Hence, the challenge to find solutions that improve socio-ecological value for those places without requiring large investments remains.

In the first presented case study, the chosen strategy to enhance residual spaces relies on applying 'minimal interventions' (Lassus 1998; Oskam et al. 2021) that stimulate both biodiversity and social accessibility. The proposed interventions resemble miniature planets, as they are roughly spherical in shape and have differentiated interiors (Schmidt et al. 2007). These 'planetoids' are 0.5–1.0 m diameter artefacts (Fig. 4.1) large enough to relate to the architecture of the site and small enough to be easily handled by humans. Their interior porosity contributes to the development of ecosystems by hosting various species (Oskam et al. 2021) either

H. Bier (✉) · A. Hidding · M. Latour · P. Oskam · H. Alavi · A. Külekci
Faculty of Architecture and the Built Environment, Delft University of Technology (TU Delft), Delft, Netherlands
e-mail: H.H.Bier@tudelft.nl

P. Morel and H. Bier (eds.), *Disruptive Technologies: The Convergence of New Paradigms in Architecture*, Springer Series in Adaptive Environments,
https://doi.org/10.1007/978-3-031-14160-7_4

Fig. 4.1 Minimal interventions in residual spaces have the potential to stimulate biodiversity and social accessibility

growing from earth balls[1] with plant seeds placed in those cavities, or from colonizing insects and animals. The goal is to mitigate the negative effects of climate change on a local level involving biodiversity loss and urban heat islands (Selwood and Zimmer 2020) by introducing microclimatic heterogeneity that buffers species against local extirpations (inter al. Suggitt et al. 2018).

In order for bypassers to interact with the plant growth and animal colonization processes, the 'planetoids' contain sensors that identify the location, temperature, humidity, etc. Data is recorded and shared via the Internet, where changes detected in the 'planetoids' are visualized and made accessible to potential visitors. The sensors may, for instance, 'indicate' that the soil of the plants is too dry, thus 'inviting' visitors to water them (Oskam et al. 2021) or pick them up and move them to locations that have less sun or are better protected from the wind.

4.2 Design-to-Robotic-Production and -Operation

Numerous applications involving Cyber-physical Systems (CpS) and Artificial Intelligence (AI) are being increasingly employed (inter al. Rajkumar et al. 2010) in architecture. Furniture scale examples such as The Big Data (2016)[2] furniture and Media Block Chair (2012)[3] are relevant examples wherein the furniture reacts to people's movement by changing colour. Such furniture takes advantage of location-based context-aware services and Internet connectivity that are ubiquitously available. It has an embedded intelligent system able to connect with, anticipate, and respond to users' desires by utilizing a variety of sensors and actuators located inside the system.

Big Data furniture, for instance, analyzes its surroundings and communicates with users by changing colours in response to movement and changing spatial factors. It

[1] Earth balls consist of a variety of seeds integrated into balls of clay, humus and/or compost.

[2] Link to Bassala website: sn.pub/zif69e and sn.pub/MJOW31.

[3] Link to TL website: sn.pub/nAwQ17.

communicates with users and other smart devices in the same room to influence their behaviour via a Twitter account. The movement and behaviour of users as well as environmental data patterns are shown in an online database. Data collected over time on the distance of users to the furniture or information on temperature, air quality, humidity, and light intensity is used to improve users' experience.

At urban furniture scale, Data-Space (van Ameijde 2019) uses a field of nodes that each incorporate a sensor and LED illumination to monitor and communicate with people within the site. The nodes are suspended above the ground in a gridded field, forming a virtual ceiling or canopy with infrared sensors to provide a real-time data stream of user locations. Movement patterns are examined and incorporated in dynamic lighting patterns that are exhibited around the visitors using a variety of evaluation algorithms and criteria aiming at motivating users to walk along light paths or encouraging physical proximity or distance between visitors. When there are too many people on the site at once, the system displays 'angry' ripple patterns giving incentives to leave.

In contrast, Flora Robotica is an automated urban garden[4] in which robots and plants perform symbiotic interactions and collaborate on the development of self-growing bio-cyber-physical structures. These structures are created using robotic controllers and mechatronic nodes with sensors and high-power LEDs that control the growth of natural climbing plants. The robots are equipped with sensors that inform them if a plant is growing nearby. They can then communicate this information amongst themselves to orchestrate the emission of lights and control the formation. Environmental impacts of the plants can be read in real time, which gives the possibility to create loop systems and to train the plants through feedback (Wahby et al. 2018a, b).

As a small-scale interactive urban furniture, the 'planetoid' relies on a methodological framework developed in 2014 in the Robotic Building (RB) lab at Technical University (TU) Delft, D2RP&O, which aims to integrate Cyber-physical Systems (CpS) into buildings and building processes (Bier et al. 2018). The goal is to link virtual and physical worlds in order to extend human capabilities and improve human and non-human interactions.

D2RP is implemented by means of parametric design and robotic production involving 3D printing with wood-based biopolymers, while D2RO techniques are implemented for the integration of sensors–actuators in order to track microclimates within and around the 'planetoid'. Data is then streamed to an app, on which users/visitors can read the real-time data and choose to interact with the 'planetoids' and their microclimates by, for instance, irrigating the plants or just playfully interacting with their light- and/or sound-based actuators (Figs. 4.1, 4.2, and 4.3).

[4] Link to FR website: sn.pub/71rCfE.

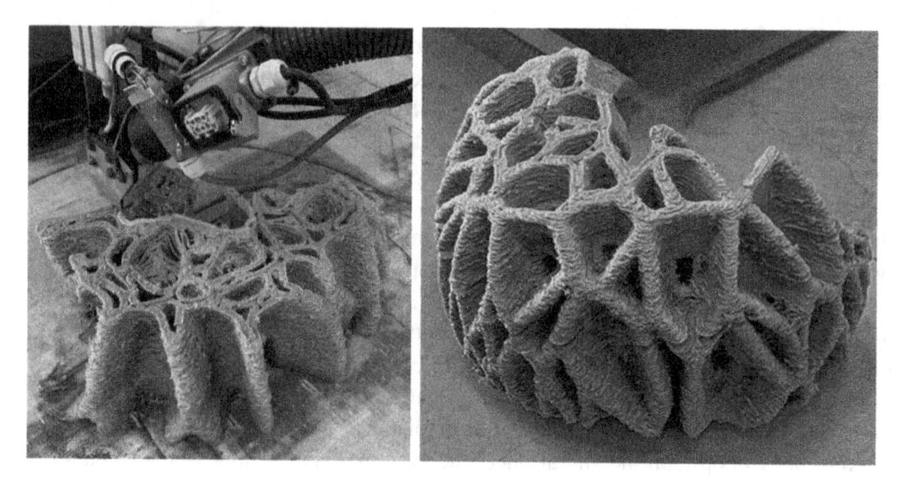

Fig. 4.2 D2RP process (left) and robotically 3D printed fragment of the 'planetoid' (right)

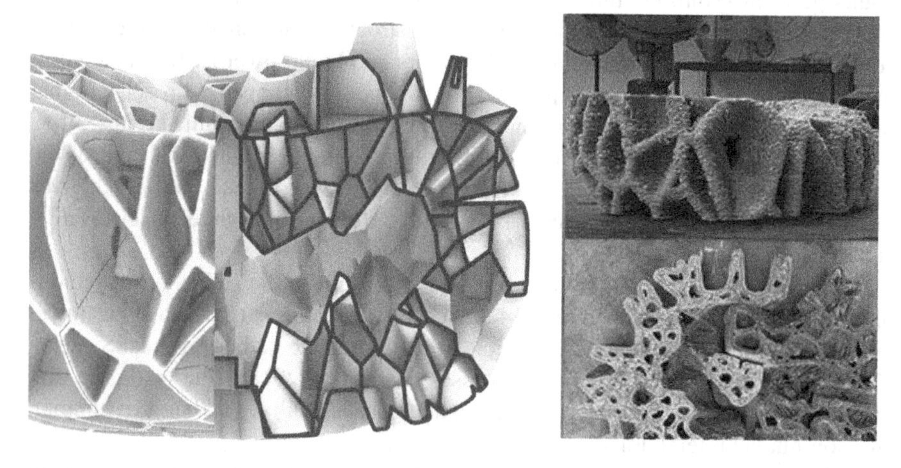

Fig. 4.3 The Voronoi structure facilitates the creation of convex and concave areas that offer opportunities for catching or repelling sun and rain and foster animal and plant species as well as sensors–actuators

D2RP

The overall shape is informed by the various functionalities of the 'planetoid' from hosting plants, insects, and small animals to harbouring sensors–actuators for monitoring the environment and communicating with visitors. These functionalities require a material design that accommodates variable porosity while catering to environmental and structural requirements. Hence, an adaptive Voronoi mesh approach is adopted (Fig. 4.3) and various aspects from function, form, material, component, and materialization are considered.

Functional Layout and Form-Finding

The overall form, porosity, and surface tectonics are informed by the use, structural requirements, and environmental conditions of the 'planetoid'. Structural forces, shadows, solar radiation, as well as rain data, are mapped onto the overall basic geometry, which is generated with the main use in mind, to protect an earthy ball with seeds. If the seeds require more exposure to the sun, there are larger openings on the parts of the shell that have the highest solar radiation. If the seeds need to be protected from direct sunlight, the openings are smaller, while the rainwater is guided either away from or towards the seeds depending on the type of seeds and the respective environmental conditions. In this context, solar radiation is calculated for the location of the object as well as the relevant time of the year, i.e., blooming period. For that purpose, Energy Plus Weather Files (EPWF) for the chosen location in Rotterdam were employed from the Climate website.[5] The data was imported into Ladybug, which is a Grasshopper plugin, in order to calculate the solar radiation.

Material Design

The Voronoi mesh is robotically 3D printed using a biopolymer consisting of cellulose, hemicelluloses, and lignin, which is processed from sawdust that is mixed with a binder, in this case, a thermoplastic elastomer (TPE). The use of biopolymers is of particular interest because of their potential to promote sustainable approaches by reducing environmental footprints (inter al. Correa et al. 2015; Kariz et al. 2016). Considering that the CO_2 released when they degrade can be reabsorbed by trees grown to replace them, biopolymers are close to being carbon neutral.

Support free 3D printing is achieved by controlling the angles of the Voronoi cells to be within the printing limitations. The maximum achievable printing angle depends on the viscosity of the material at extrusion temperature as well as cooling, i.e., crystallization speed. The printing angles are limited to 45–55 degrees in relation to the printing bed. Since the Voronoi cellular structure is an inherently stable self-supporting type of geometry, the cells can be printed at more extreme angles. The overall geometry is subdivided into Voronoi cells that enable the control of global and local porosity. The 'planetoids' are more porous in some areas than in others in order to accommodate structural and environmental requirements and most importantly programmatic requirements for plants, insects, and small animals of various sizes as well as sensors-actuators (Fig. 4.3). The overall goal is to have multiple performances addressed with a consistent material design at macro, meso, and micro scales.

[5] Climate website: sn.pub/aP7d3n.

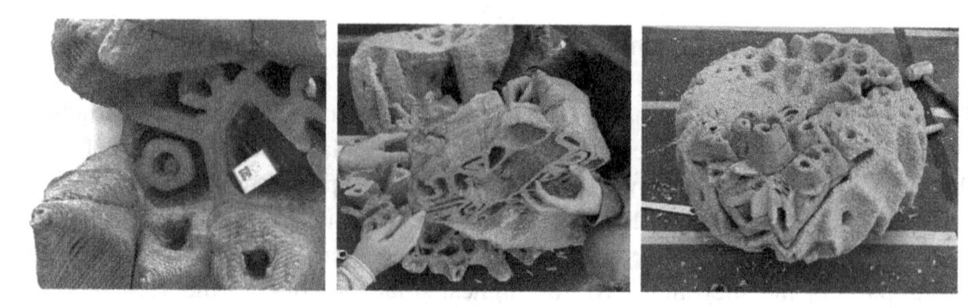

Fig. 4.4 Integration of sensors and batteries (white boxes) in component (left), componential logic (middle), and assembled prototype

Componential Logic

The prototype was subdivided into multiple components, allowing the 'planetoid' to be printed in multiple parts. Based on this strategy larger objects can be created out of multiple components. The total size of the assembled object then is not limited to the size of the 3D printing system (Fig. 4.4). Also, easy transportation and assembly are accounted for.

Tool Paths

Continuous toolpaths ensure that the printing process is efficient. The production time is only defined by the object size, layer height, and speed of the 3D printer. With a layer height of 2.0 mm and a printing speed of around 300 mm per second, the process took about 20 hours. It was important to optimize the tool path and eliminate travel moves because, at the start and end points of the travel moves, the 3D printer has to stop and start printing. Every starting and stopping location leaves a mark in the 3d print, so it is best to minimize these starting and stopping moments. Hence, the continuous toolpaths that were generated ensured efficient production time and improved quality.

Prototype

While the D2RP part has already been completed (Figs. 4.1 and 4.2), the D2RO work is still in progress. Multiple sensors–actuators are integrated into the 'planetoid' in order to monitor microclimates and initiate activities. Data is streamed to an app, on which users/potential visitors are notified in real time and are 'invited' to interact with the 'planetoids' and their microclimates.

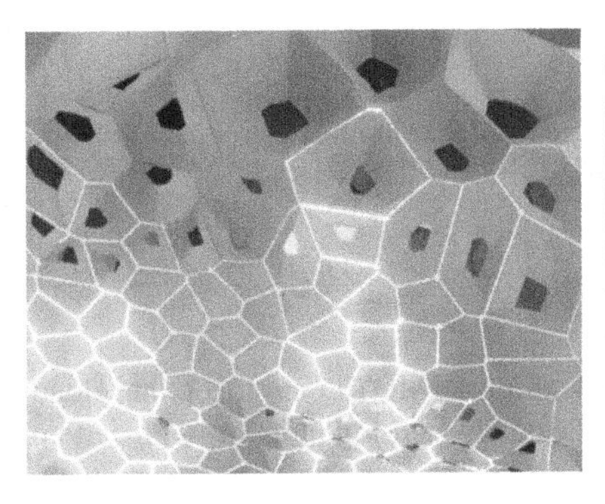

Fig. 4.5 Sensors–actuators proximity sensors and light actuators (left) and plants (right) integrated into the Voronoi structure

D2RO

The 'planetoid' offers a protected environment for hosting earth balls with seeds that develop into plants as well as animals and sensors-actuators. If the natural systems consist of (i) plants such as dandelions, camomile, and poppies, (ii) insects such as butterflies, dragonflies, and bees, and (iii) small animals such as snails, hedgehogs, and rodents, the integrated sensor–actuator system consists of various components that require further definition.

Sensing Modules

Each sensing module (Fig. 4.5) carries a unique identifier, defining its function as well as modes of functioning including frequency of data collection and communication. Each 'planetoid' hosts several sensing modules which, independent of the others, can be added, maintained, and modified. The sensors require a remarkably low amount of energy and can operate on a battery for several months.

Gateway

One gateway collects, via Bluetooth, the data transmitted by all the sensing modules in its physical proximity. It broadcasts the sensor data along with the identifier of the sensing module to the LTE urban antenna. The transmitted data also contains information about the cloud service associated with this setup as well as the credentials to access the cloud database.

Network

The LTE (Long-Term Evolution) networks are available in most European cities by various commercial providers and will be used.

Communication

Through the Message Queue Telemetry Transport (MQTT) protocol the cloud database 'subscribes' to receive the data collected by sensing modules with certain identifiers. The data will be stored and made available for queries through any web-based application. Data is made accessible to the users through either the Quick Response (QR) code associated with the 'planetoid', or simply its placement. Gamified presentation of data aims to be engaging and leading to action (Figs. 4.6 and 4.7).

Various sensors concerned with monitoring humidity, light, temperature, and the presence of humans and actuators involving light and sound are integrated into the 'planetoids'. In its future development, the 'planetoid' will rely on learning capacities to predict moments—depending on the patterns of human and non-human activities around the planetoid—when opportunities arise for interaction with the

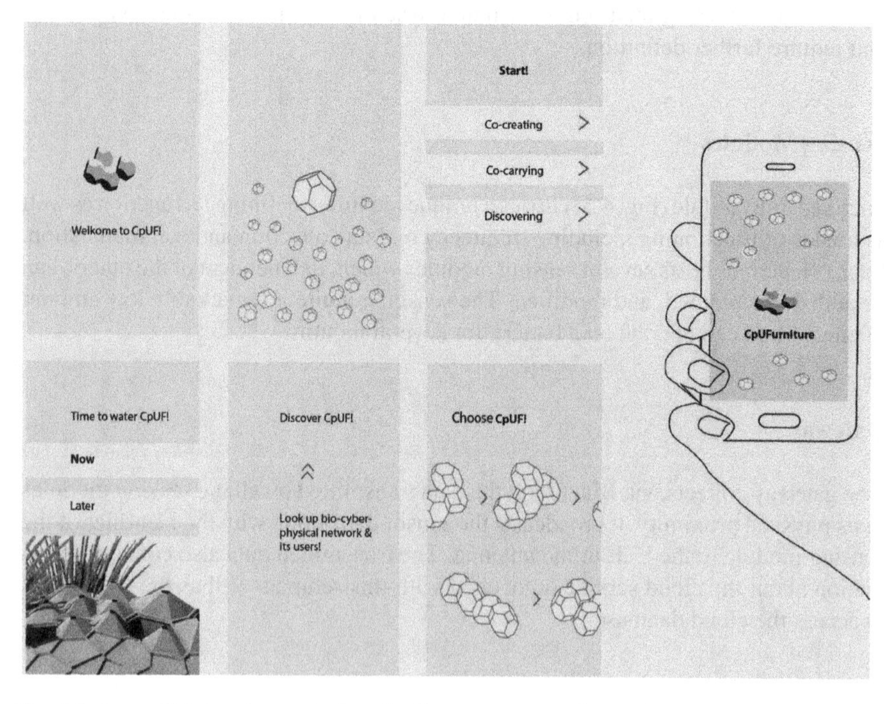

Fig. 4.6 App with interaction modalities for discovering, co-caring, and co-creating environments populated with 'planetoids'

Fig. 4.7 Urban furniture accommodating human activities such as climbing, sitting, and lying down developed with students

evolving nature (vegetation, insects, etc.) and humans. K-means and Hierarchical Clustering (HC) as established Artificial Intelligence (AI) methods will be applied to discern correlations between presence, movement, and actions with weather variables, to be able to offer a structured prediction of opportune interaction moments and to promote them through an open access mobile application.

AI has been used in the built environment for applications involving transportation networks, water, lighting, and heating systems. It also has been increasingly employed to provide safer public spaces and services (inter al. Cugurullo 2020; Chew et al. 2021). The interaction scenarios for the 'planetoids' engage users in discovering, co-caring, and co-creating (Fig. 4.6) by involving users in learning about plants, animals and their needs, encouraging them to water, weed, plant new seeds, depending on the monitored development reconfiguring 'planetoids' by moving or adding more 'planetoids', etc.

Interaction Scenarios

The interaction scenarios involve activities such as (a) monitoring plants, insects, and animals and (b) involving users. The goal is to engage neighbours and passers-by with the 'planetoids' and their environments that go through several stages of development and transformation from bare to by plants overgrown planetoids. By employing real-time sensing of the natural and human activities around the planetoid such as passing by, sitting, lying down, etc., users are made aware of various forms of life and engage with them in interactive experiences that can reap some of the potentials of abandoned areas as public urban spaces. The system is sensing environmental parameters such as temperature, humidity, and light, as well as information related to the presence and movements of humans, animals, and insects around the 'planetoids'. The main actuation is in the form of mobile application (app) notifications informing the users about the emerging activities around the planetoids or the need for their action (e.g., watering the plants).

Additional actuation is envisioned as serving educational, playful, and cautionary purposes. All three rely on outputs such as text, charts, images, movies, steamed videos, lights, and sounds as elements of the sensor–actuator system aiming to respond to environmental and human input. In this context, three scenarios are envisioned:

(a) Educational: By locating 'planetoids' within the urban context on the app and by learning about their microclimates users develop awareness and may increasingly engage with the 'planetoids' to help them thrive.
(b) Cautionary: By notifying users via text, light, and sound when humidity in the 'planetoid' is low or other hazards including vandalism, the 'planetoid' engages users in a 'supportive' relationship.
(c) Playful: By turning on the integrated sources of lights and sounds on and off, changing their intensity and colour, etc., users become co-creators of the emerging bio-cyber-physical environment.

In this context, the app is meant to customize interaction. If the usual interaction is based on simple patterns of light and sound responding to passers-by, the app allows users to potentially co-create AI-supported music and light compositions. The most important interactions are (i) engage, (ii) co-create, and (iii) disengage: As soon as the system 'notices' movement, the lights pulsate in one colour—i.e., oscillates between intensities of the same color indicating that the 'planetoid' comes to 'life'. The intention is to instigate interest and curiosity in the users, inviting them to engage with it. When engagement is established a gradual shift from the initial colour and pulsating pattern, to changing colours and patterns that are customizable by users takes place. When more users and planetoids are engaging in interaction AI comes into play to direct and moderate the interaction by reinterpreting and recomposing music using AI Virtual Artist (Barreau 2018) or AI Duet by Google that is trained to respond to midi tones in a harmonious way. Instrument samples are combined and the resulting musical composition is dynamically visualized through light using an approach similar to the Music Animation Machine (Adli et al. 2007). The main purpose is to encourage social interaction and facilitate social gatherings. The disengagement is activated when users leave the physical and/or virtual space in which the 'planetoids' are located. In the disengaged 'dormant' state the 'planetoid' only reacts to vandalism by activating shrill sounds and lights.

The sound-light compositions change in time as the imbued AI learns from the communities interacting with the 'planetoids' that can be aggregated into groups to create harmonious spatial-sound-light compositions. Some areas may become more frequented by youth engaging in creating sound-light compositions for outdoor parties, while others may become more suitable for the elderly by contributing to the revitalization of residual spaces. In order to serve such revitalization, the AI has to monitor the well-being of all actors and adjust interaction scenarios accordingly at all times.

Scaling up Scenarios

Explorations towards scaling up the system by combining 'planetoid'-sized components into larger pieces of urban furniture (Fig. 4.7) has been initiated with consideration to human needs. The possible combinations of components are explored on the app (Fig. 4.6) in order to customize designs depending on specific needs, for instance, sitting, lying down, climbing, etc. The combinatorial logic has been investigated in studies implemented with Ph.D. and M.Sc. students in connection to several funded projects in collaboration with academic and industrial partners (Bier et al. 2021; Oskam et al. 2021). The conceptual design involved the study of activities and potential new activities in respective locations ranging from playing to lounging or doing sports. In addition, factors such as environment, pre-existing infrastructure, urban context, flora and fauna, etc., were considered. Once the parameters to inform the design were identified, the form-finding process was initiated using the driving force activity patterns and materialization by means of robotic 3D printing using biopolymers.

4.3 Discussion

Socio-technical interventions made in natural environments to improve biodiversity and human–robot interaction are not new. Various projects involving artificial reefs and 3D printed scaffolding for microorganisms (inter al. Gautier-Debernardi et al. 2017) have shown that eco-friendly solutions can meet the needs for increasing biodiversity in various natural environments. Also, projects involving sensor–actuator networks such as Data-Space (van Ameijde 2019) using a field of nodes that each incorporate a sensor and LED illumination to monitor and communicate with people within the site prove their potential to engage humans in various interactions.

The 'planetoids' described in this chapter act as socio-technical interventions that not only improve biodiversity, but also increase human–nature interaction as well as social accessibility of leftover spaces by employing sensor–actuator networks. These allow monitoring of development in a time of newly established habitats on the 'bio-cyber-physical planetoid' app that is inviting potential visitors to irrigate the 'planetoids' or protect them from the sun, or playfully interact with them and with each other.

The novel opportunities offered by cybernetic social-ecological systems involving AI and their ability to identify in this case correlations between the evolving nature, weather variables, and actions of humans in order to offer a structured prediction of opportune interaction moments and to promote them through open access web-based platforms and mobile applications establish bio-cyber-physical feedback loops that render human and non-human agents as co-creators of processes and events.[6]

[6] Cyber-physical Space and Urban Furniture wikis: sn.pub/uyTFMl and sn.pub/YQrlk8.

Acknowledgements This project has been partially funded by the Dutch Research Council, the Creative Industries Fund NL and FCT Portugal—'Fundação para a Ciência e Tecnologia' and has profited from the contribution of the 'Microruin Lab', ID+ Research Institute for Design, Media and Culture. It has been co-funded by PoliMi and UniFri. The D2RP&O process has been developed in the Robotic Building lab by researchers and students participating in the 'Bio-cyber-physical Planetoids' and 'Cyber-physical Urban Furniture' projects. The robotic 3D printing has been implemented with the system of 3D Robot Printing and the sensor–actuator system has been provided by Starnberger Innovation and Technology and has been integrated with help of UniFri.

References

Accordino J, Johnson GT (2000) Addressing the vacant and abandoned property problem. J Urban Aff 22(3):301–315. https://doi.org/10.1111/0735-2166.00058

Adli Z, Nakao, Nagata Y (2007) A content dependent visualization system for symbolic representation of piano stream. In: Knowledge-based intelligent information and engineering systems: 11th international conference, KES 2007, proceedings. Springer. p 292. ISBN 9783540748267

Barreau P (2018) How AI could compose a personalized soundtrack to your life. TED 2018: the age of amazement. https://link.springer.com/chapter/10.1007/978-1-4939-2836-1_11. Accessed 30 Sep 2021

Bier H, Cheng AL, Mostafavi S, Anton A, Bodea S (2018) Robotic building as integration of design-to-robotic-production and-operation. In: Robotic building. Springer, pp 97–119

Chew L, Hespanhol L, Loke L (2021) To play and to be played: exploring the design of urban machines for playful placemaking. Front Comput Sci 3:635949

Correa D, Papadopoulou A, Guberan C, Jhaveri N, Reichert S, Menges A, Tibbits S (2015) 3D-printed wood: programming hygroscopic material transformations. 3D Print Addit Manuf 2(3):106–116. https://doi.org/10.1089/3dp.2015.0022

Cugurullo F (2020) Urban artificial intelligence: from automation to autonomy in the smart city. Front Sustain Cities 2:38. https://doi.org/10.3389/frsc.2020.00038

Edensor T (2005) Industrial ruins: space, aesthetics, and materiality. Bloomsbury Academic

Gautier-Debernardi J, Francour P, Riera E, Dini E (2017) The 3D-printed artificial reefs, a modern tool to restore habitats in marine protected areas. The Larvotto-Monaco context. In: Proceedings of international marine protected areas congress Chile 2017

Haase A, Bernt M, Großmann K, Mykhnenko V, Rink D (2016) Varieties of shrinkage in European cities. Eur Urban Reg Stud. https://doi.org/10.1177/0969776413481985

Harrison C, Davies G (2002) Conserving biodiversity that matters: practitioners' perspectives on brownfield development and urban nature conservation in London. J Environ Manag 65(1):95–108

Kariz M, Sernek M, Kuzman MK (2016) Use of wood powder and adhesive as a mixture for 3D printing. Eur J Wood Prod 74:123–126. https://doi.org/10.1007/s00107-015-0987-9

Kawata Y (2014) Need for sustainability and coexistence with wildlife in a compact city. Int J Environ Sci Dev 5(4):357

Kowarik I (2013) Cities and wilderness. Int J Wild 19(3)

Lassus B (1998) The landscape approach. University of Pennsylvania Press

Laurie IC (1979) Nature in cities: the natural environment in the design and development of urban green space LK. https://tudelft.on.worldcat.org/oclc/3361580

Oskam PY, Bier H, Alavi H (2021) Bio-cyber-physical 'planetoids' for repopulating residual spaces. Spool CpA TU Delft

Rajkumar R, Lee I, Sha L, Stankovic J (2010) Cyber-physical systems: the next computing revolution. In: Proceedings—Design automation conference, pp 731–736. https://doi.org/10.1145/1837274.1837461

Schmidt B, Russell CT, Bauer JM, Li J, McFadden LA, Mutchler M, et al (2007) Hubble space telescope observations of 2 pallas. Bulletin of the American Astronomical Society
Schwarz U (1980) Der Naturgarten: mehr Platz für einheimische Pflanzen und Tiere
Selwood KE, Zimmer HC (2020) Refuges for biodiversity conservation: a review of the evidence. Biol Conserv 245. https://doi.org/10.1016/J.BIOCON.2020.108502
Suggitt AJ, Wilson RJ, Isaac NJB, Beale CM, Auffret AG, August T, Bennie JJ, Crick HQP, Duffield S, Fox R, Hopkins JJ, Macgregor NA, Morecroft MD, Walker KJ, Maclean IMD (2018) Extinction risk from climate change is reduced by microclimatic buffering. Nat Climate Change 8(8):713–717. https://doi.org/10.1038/s41558-018-0231-9
van Ameijde J (2019) The architecture machine revisited: experiments exploring computational design-and- build strategies based on participation. SPOOL 6(1):17–34
Wahby M, Heinrich MK, Hofstadler DN, Neufeld E, Kuksin I, Zahadat P, Schmickl T, Ayres P, Hamann H (2018a) Autonomously shaping natural climbing plants: a bio-hybrid approach. R Soc Open Sci 5:180296. https://doi.org/10.1098/rsos.180296
Wahby M, Heinrich MK, Hofstadler D, Zahadat P, Risi S, Ayres P, Schmickl T, Hamann H (2018b) A robot to shape your natural plant: the machine learning approach to model and control bio-hybrid systems. https://doi.org/10.1145/3205455.3205516

Chapter 5
Data-Driven Urban Design: Conceptual and Methodological Constructs for People-Oriented Public Spaces

Jeroen van Ameijde

5.1 Introduction

The increasing integration of information and communication technologies into urban spaces allows cities to become sensing systems, offering city planners and managers a fast-expanding spectrum of data that reveals the inner workings of our urban environments. A wide range of 'smart city' and urban research applications has emerged around these opportunities, using urban analytics, big data analysis and urban modelling to evaluate urban processes based on real-world or even real-time information. These initiatives can be grouped within the emerging field of *urban informatics*, which has been defined as "an interdisciplinary approach to understanding, managing, and designing the city using systematic theories and methods based on new information technologies, and grounded in contemporary developments of computers and communications" (Shi et al. 2021, p. 1).

The increasing availability of urban data could help reduce the gap between the social sciences and urban planning practice (Dyer et al. 2017), creating pathways to connect quantified insights into urban processes to more precisely calibrated and refined urban design proposals. As many types of urban data relate to human activities, strategies for data-driven urban design informed by urban monitoring and analysis have the capacity to be more people-oriented, location-specific and diverse, compared to traditional urban design methods which are often based on assumptions and generalised 'best practice' methods. This could imply an increased focus towards an end-user-oriented approach, which is a significant shift in a practice that often prioritises commercial rather than community interests (Nisha and Nelson 2012).

The data-driven connection between urban sensing and management decisions raises questions that are bound to increase in importance as the scale and scope of

J. van Ameijde (✉)
The Chinese University of Hong Kong, Hong Kong, China
e-mail: jeroen.vanameijde@cuhk.edu.hk

P. Morel and H. Bier (eds.), *Disruptive Technologies: The Convergence of New Paradigms in Architecture*, Springer Series in Adaptive Environments,
https://doi.org/10.1007/978-3-031-14160-7_5

digital integration are set to expand significantly in the coming years. These questions focus on issues such as transparency of decision-making, ownership of data and control over how urban data is interpreted (Leszczynski 2015; Mann et al. 2020). Urban scholars have pointed towards positive opportunities, as well as potential negative consequences of the 'apparatisation' (Lee and Bier 2019) of urban environments. Data-driven urban design could forefront human-centric and evidence-based methods to create inclusive neighbourhoods designed through collaborative or participatory processes (van Asselt and Rijkens-Klomp 2002; Foth et al. 2018). It could also increase the influence of corporations that may prioritise financial objectives rather than citizens' concerns (Tombs and Whyte 2015), giving rise to a type of smart city that uses mass-data gathering to reinforce existing "systemic injustices, biases, inequalities, and power structures" (Mann et al. 2020, p. 1104).

The positivist approach towards the digitisation of urban environments focuses on the potential for increased citizen participation, as sensing systems turn cities into "a reflexive test-bed and workshop for connected habitation in enmeshed digital and physical space" (Ratti and Claudel 2016, p. 23). These feedback loops can operate across different scales of urban space and across a range of time intervals. Similar to how building interiors are increasingly managed through digital signage, climate control and security systems, urban spaces can be controlled in response to people's location data, movement and activity patterns. New urban infrastructure and services can respond to real-time demand based on mobile geolocation and communication protocols, creating a layer of digital services and relationships superimposed on the physical city. Similar to Christopher Alexander, who saw an opportunity in early computer systems to enable "a shift from abstract, overly intellectualised design to an approach based on people's immediate daily needs" (Gehl and Svarre 2013, p. 53), technology could be employed to empower people, allowing them to take ownership of urban areas and influence improvements based on local needs.

Several researchers have, in recent years, investigated how new paradigms of feedback and control could be implemented through technologically enhanced built environments. In their book "The city of tomorrow: Sensors, networks, hackers, and the future of urban life" (2016), Ratti and Claudel present a range of projects based on sensing, interpreting and influencing environmental processes, movements and activities within cities. Their vision of augmented urban life is based on open data and platforms to enable grassroots initiatives and non-profit models of urban co-creation (Ratti and Claudel 2016). Lee and Bier (2019) explore how the scope of the design propositions should be expanded to address the 'apparatisation' of architecture. The conceptualisation of spaces, interfaces and protocols for environments that accommodate dynamic social processes, will require "the hybridization of disciplines such as architecture, interaction design, sociology, psychology, biology, and computer sciences (…), which must overcome the limits and constraints of the disciplinary territories" (Pillan et al. 2020, p. 54). Researchers at TU Delft developed a series of 'Cyber-physical System'-based projects, to test embedded sensing systems, designed to establish intelligent relationships with human activities. The 'Omnipresence' project incorporated Machine Learning to collect data from users and the environment and learn to respond to users' needs over a longer period of time (Pavlovic et al. 2020). The project 'Connected Lighting for a Caring City' used interactive lighting to assist the elderly in daily activities and improve their quality of life. The projects were part of a larger research agenda that explores how adaptive environments can be developed as sensitive, transparent and democratic support systems (Pillan et al. 2020).

The wide range of recently emerged 'smart city' technologies might seem to enable citizen empowerment and societal progress, as smart city projects are often promoted as means to increase "prosperity, sustainability, and liveability on the basis of a presumed technological neutrality" (Mann et al. 2020, p. 1104). A broad group of urban scholars, however, has explored how in many applications, smart city technologies reinforce systems of centralised control, as data is monopolised by governmental or commercial organisations (Kitchin 2017; Miller 2018). Smart city projects expand private sector interests into the public domain (Koolhaas 2014), and there are growing concerns around the repurposing of social media, smartphone and sensing data without citizens' informed consent (Pierson 2012). Smart city technologies operated by corporate providers may be "forging a new social contract on societies" (van Dijck 2014, p. 206), as automated mass surveillance turns citizens into involuntary providers of monetizable data (Barns 2020; van Doorn 2018).

Scholars have argued that smart city technologies may have a profound effect on the freedoms, inclusivity and sense of participation experienced in future public spaces, and that Lefebvre's notion of 'the right to the city' (1968) is under threat by processes of privatisation and monopolisation, propelled by the neoliberal management policies of cities in both physical and digital strata (Kitchin 2015; Vanolo 2014). As the use of digital services is becoming unavoidable to participate in public urban life, the accessibility of smart city spaces is limited by unequal access to devices and operating systems by vulnerable groups, and under threat of increased censorship and control within online worlds (Foth et al. 2016; Shaw and Graham 2017). Critical scholars have argued for a new paradigm of governance relating to smart cities, in which ownership rights, civil and social participation are protected by a set of governing principles based on normative rather than neoliberal ideals (Breuer and Pierson 2021; Cardullo 2020). De Lange (2019) expands upon the notion of the 'urban data commons' to prioritise a humanist view of smart cities over a cybernetic one and promote the coming together of data, human actors and urban issues to enable inclusive citizen participation in the digitised city.

Ratti and Claudel are reservedly positive about technology companies' offering of free services in exchange for the commodification of user data, as their urban-scale digital networks could deliver "ecosystems of technology, assimilated in urban space" that "derive maximum resource efficiency by working coherently and systematically" (Adam Greenfeld, quoted in Ratti and Claudel 2016, p. 31). Other scholars have provided extensive critical analyses on the topic of 'platform urbanism', the emerging spatial significance of connected citizens through social media and smartphone applications, which serve as platform ecosystems to connect providers and consumers of services at scale. Barns (2020) has explored how on one hand, these platforms seem to enable "'collaborative' forms of consumerism and peer-to-peer exchange" (Barns 2020, p. 81), which has led to their mass adoption and their association with "urban innovation and progress in a digital age" (ibid. p. 82). On the other hand, she argues, people were mostly unaware of the benefits for the platform owners, who turned activity data into "training data to support the algorithmic governance of user behaviour" (ibid. p. 196). While a recently increased public awareness has led to more governmental scrutiny and regulation around 'Big Tech' companies' operations, a continuing debate on these aspects will need to shape a future balancing act between people's "surveillant anxiety" (Crawford 2014) and their desire or dependencies on the benefits created by the platforms. The initial promise of digital platforms to enable a "worldwide, 'bottom up' disruptive urban movement of social change, taking place outside of established modes of exchange and institutional organisation" (Barns 2020, p. 196) has given way to a different reality, where the newly emergence mechanisms of digital control require a rethinking of the possibilities for stakeholder participation and data management policy debates.

The literature on smart cities and digital urbanism shows how data-driven urban design processes are already shaping current and future cities, outside the scope of architects' and urban designers' mode of practice and are often driven by economic development objectives rather than social policy initiatives. While a broad range of critical scholarship has offered detailed analysis and pathways of resistance to the potential negative incarnations of digital urbanism, there is a need to explore the conceptual mechanisms and policies that architects and planners could develop to implement positive design applications within this paradigm. To bridge the gap between urban scholarship and practice, there is a need to investigate and conceptualise the methods and tools of data-driven design, to provide insights and guidelines for future research and urban design applications, operating within tomorrow's technology-driven cities.

This chapter presents a series of theoretical and procedural experiments that explore potential scenarios and implementation mechanisms around the data-driven integration of urban analytics and generative design. To frame these projects within a historic-conceptual context, the chapter starts with an analysis of the speculative participatory urban design processes developed in the 1970s by Negroponte, Friedman and Pask. These early explorations of cybernetic systems are then connected to the contemporary theoretical framework of placemaking, which focuses on environmental and social behaviours, and relationships. Subsequently, a series of on-site experiments are discussed in which the principles of placemaking are quantified and structured into computational generative design processes. Lastly, current research developments are discussed in which the notions of data-driven research into placemaking are interpreted within the context of applied research and urban design implementation scenarios.

5.2 The Architecture Machine

The idea of integrating physical and digital layers in the built environment has been conceptualised since the emergence of computing systems in the early 1960s when pioneers such as Christopher Alexander, Richard Saul Wurman, Cedric Price and William Mitchell speculated on their capacity to understand and generate organisational structures of architectural and urban spaces. Gordon Pasks' 1969 article "The Architectural Relevance of Cybernetics" powerfully promoted the notion of architecture becoming "a mechanism of information exchange", and for architects to adopt the role of 'system designers' (Steenson 2017, p. 17). Nicholas Negroponte, and his 'Architecture Machine Group', founded at MIT in 1967, speculated on systems that could democratise and localise control over the design of the built environment, developing 'humanistic' machines that could respond to user requirements, analyse user behaviour, and anticipate future problems and solutions (Negroponte 1970). Negroponte, together with colleagues and students, developed working prototypes of interactive systems, breaking down the complex mechanism of monitoring, analysing and actuating user-based environments into rule-based systems that govern responses.

The conceptual notion of 'The Architecture Machine' describes an abstract machine, theorised by others as a device that "does not function to represent, even something real, but rather constructs a real that is yet to come" (Deleuze and Guattari 1987, p. 142). Explored in their lab, established at MIT in 1967, and in books published in 1970 and 1975, Negroponte and his collaborators set out a utopian vision that combines humanism with futurism, speculating on the possibility for combined man–machine processes to translate client requirements and desires accurately into a detailed architectural design, without the interference of the architect's self-interests:

> In most cases the architect is an unnecessary and cumbersome (and even detrimental) middleman between individual, constantly changing needs and the continuous incorporation of these needs into the built environment. The architect's primary functions, I propose, will be served well and served best by computers (Negroponte 1975, p. 1).

The most well-known project produced by the Architecture Machine Group was an art installation titled 'SEEK', shown during the 'Software' exhibition in New York in 1970. The installation consisted of a large plexiglass enclosure containing a three-dimensional landscape of small cubes, and a number of small mouse-like animals whose movements would disrupt the cubes. A computational feedback loop consisting of a camera and a robotic arm was calibrated to analyse and amplify changes made by the gerbils. The ideal 'final' configuration of the cubes would emerge over time, out of the interaction between the inhabitants and their architectural environment. The 'SEEK' project was technically considered a failure as there was no stable outcome achieved by the system, and it was criticised for "inappropriate abstraction of real-world constraints and too great a scope of the design problem at hand" (Steenson 2017, p. 184). Yet, Negroponte's experiments made a provocative and influential contribution to the discourse around participatory urban design, introducing the notion of end-user agency, and of responsive environments based on human–machine interaction systems programmed according to certain policies. It demonstrated the ambition to "bring urban design back the ordinary man" (Rowe 1972, p. 12), allowing citizens to be involved in the complex negotiations around urban problems through mediation by a fair and transparent system containing computational rules.

The larger agenda of Negroponte and his group was to develop methods for more efficient and intuitive control of complex data-processing tasks, for instance, to open up design and management workflows to stakeholders and end-users. To achieve this, they worked on tools for the management of data in graphical and spatial ways, rather than through textual or numeric systems that are inaccessible to non-specialist users. In conversation with collaborators Gordon Pask and Yona Friedman, they explored prototypical software protocols for 'computer-aided participatory design' which aimed to capture end-users' 'intentionalities'. The software would in these scenarios take on the role of an intelligent partner in the design process, helping their human counterparts with collecting, processing and interpreting complex data. The notion of 'intentionality' described the necessity for software programs to distinguish between different hierarchy levels of information, analysing and abstracting data into information that captures meaning. In order for a software application to serve as a helpful resource in the design process, it needed to be able to process design information in the same graphic and spatial languages that are important for the designers and users of architectural spaces:

"(1) We want our machine partners to have the potential of perceiving those aspects of the physical environment that would become biased or incomplete when transmitted through other modes (such as a verbal description). (2) We want machines to be able to solicit information directly from the real world on the initiative of internal computations rather than depend upon the intervention of a human designer and his conscious or subconscious interpretations of that information. (3) We want computers to be able to witness and handle concepts and relationships (and even experiences) that are concerned with those environmental qualities that human designers understand and handle through metaphors and symbols" (Gordon Pask, in Negroponte 1975, p. 48).

These statements by Gordon Pask, made in relation to his larger body of work around cybernetic systems, language systems and human–machine collaboration, indicate the desire for 'architecture machines' to be in direct contact with the real world, communicating fluently with the environment that they are designed to enhance. Yona Friedman described a similar scenario of participatory design as "non-paternalist" (Negroponte 1975, p. 96), as his system was designed to not cast judgement on its users or the outcomes, but to help them produce results within certain acceptable constraints. Figure 5.1 shows Friedman's conceptual organisation, setting up relationships between design and implementation loops to assist with the negotiations between a knowledgeable user and the collective interests of a community.

Pask and Friedman's perspectives help to further interpret the 'SEEK' experiment as a cybernetic system aimed at empowering its inhabitants. Figure 5.2a describes the feedback loop of the computer-enhanced space and its separate processes of

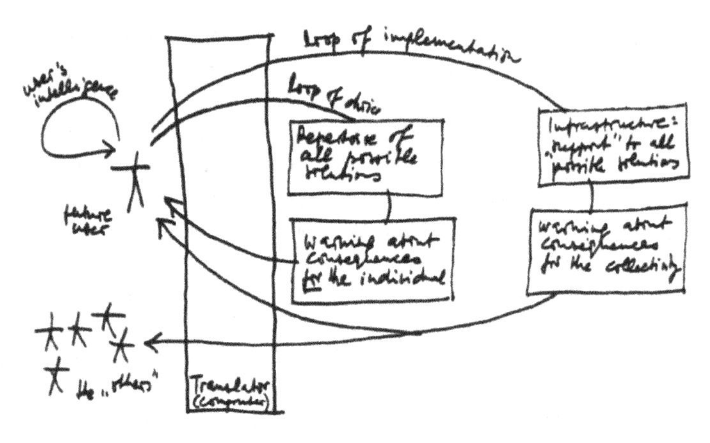

Fig. 5.1 'Computer-aided participatory design' diagram by Yona Friedman. *Source* Negroponte, N., Soft architecture machines. Cambridge, MIT Press, 1975, p. 94

scanning and processing environmental data, and of responding and implementation with environmental remediation. While this feedback loop is designed to function autonomously without the interference of designers, there is a crucial human role in defining the interpretation protocols and reactionary policies that should drive the environment's evolution. Figure 5.2b interprets this dynamic according to Friedman's 'nonpaternalist' hierarchy: denying the designer a close association with the computer system to avoid interference based on preference, but rather having the computer be part of the environment. The designer's role is to conceive, test and develop the machine protocols and machine–environment interaction, evaluating their mutual capacity to produce outcomes that satisfy all stakeholders.

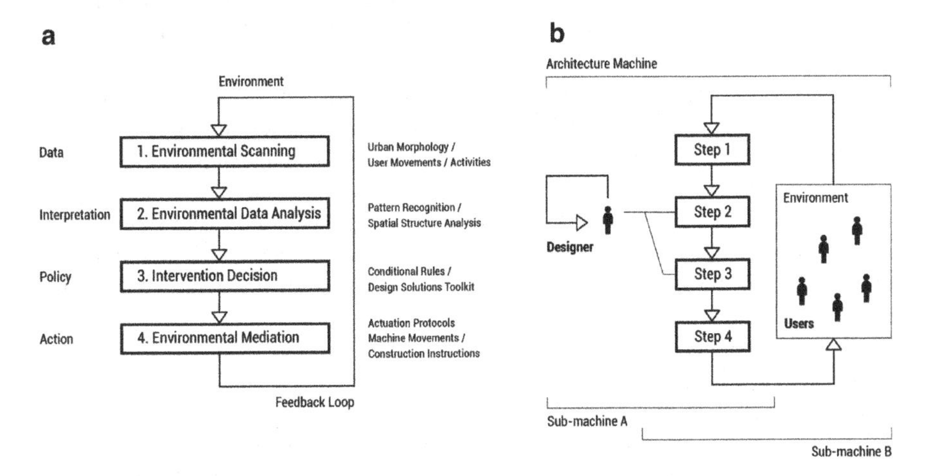

Fig. 5.2 **a** and **b** System diagrams of the 'SEEK' experiment, describing the relationships between environment, computer, and designers of the system. *Source* Author, based on Friedman

When evaluating the abstract machines explored by Negroponte and his contemporaries within the context of today's technological society, it is easy to draw parallels between these early conceptions of interactivity and participation, and the systems enabled by ubiquitous computing we see today. As primitive computing systems and clumsy robotics are replaced by powerful miniaturised and distributed devices, the urgent challenge to redefine the role and responsibilities of the urban designer persists. Negroponte's experiments aimed to provoke a debate about the ethics of creating controlled social environments, as it is crucial to reflect on the increasing influence that ubiquitous computing systems will have on the management and experience of urban spaces. As we explore how the notion of a participatory urban design process as proposed by Negroponte, Friedman and Pask can be implemented within contemporary urban design practice, we look to the notion of 'placemaking'. This concept combines a theoretical framework with practical indications on how to implement new applications of the notion of 'The Architecture Machine'.

5.3 Public Space, Urban Analytics and Placemaking

Since the 1970s, the notion of placemaking has been widely employed in the fields of social sciences and urban planning (Friedmann 2010), asserting that place is more than a location or container of human action but instead is produced by people's relations with their environment, their geographical behaviour and the social structures and identities of space and place (Tuan 1976). Cresswell (2014, p. 39) defined 'place' as "constituted through reiterative social practice", emphasising that the value of a place lies in its ability to stimulate events and social behaviours. Opposite to public space planning practice, which often promotes behavioural and economic goals, urban theorists have argued that urban environments should be designed to signify community sentiments, symbolism, identities, and psychological well-being (Stokols 1990).

The role of public space in maintaining personal and community health and well-being is increasingly acknowledged (Samuelsson et al. 2020), pointing to a growing awareness of the social significance of effective urban design solutions. As increasingly dense and costly urban developments are populated with ever smaller apartments, the shared spaces around residential buildings are used as an extension of the domestic sphere. Well-planned public spaces facilitate people's interaction with neighbours and the surrounding context, which contributes to people's well-being and integration within society (Lau and Murie 2017).

Jane Jacobs famously advocated for urban places as a concentration of 'diversity' where "one needs to learn how to live with and among strangers" (Jacobs 1961, p. 143). Besides primary social relations, Jacobs outlined how a sense of belonging and collaboration in the neighbourhood can arise from networks of 'secondary relations', cultivating a sense of trust while maintaining a sense of privacy. Research in the social sciences has shown that the sense of community is improved by resident homogeneity and length of residency, or public space circulation which causes "casual neighbouring", which can evolve into social bonding, integration and attachment to place (Talen 1999, p. 1375).

Placemaking and community forming depend on the notion of 'co-presence', as "encountering, congregating, avoiding, interacting, dwelling, conferring are not attributes of individuals, but patterns, or configurations, formed by groups or collections of people" (Hillier 1998, pp. 29–31). The patterns of co-presence in an urban environment are commonly the result of our everyday practices (Legeby 2013), which are inscribed in urban space. Social interactions and gatherings can be described through 'time geography' (Hägerstrand 1953; Yin and Shaw 2015), and the architectural structures and urban configurations of space provide the material preconditions for the patterns of movement, encounter and avoidance, acting as a generator of social relations (Hillier and Hanson 1984).

The notion of placemaking as a product of social interactions and co-presence allows for the linking of digital place studies and processes of participatory practice in urban design, as this conceptual connection provides a pathway for connecting public space analysis to strategies for the improvement of these very same spaces. While other scholars have distinguished between "natural placemaking" and "accelerated placemaking" (Foth 2017, p. 1), equating these to user-led versus developer-led community forming, a third category of 'assisted placemaking' can be conceived in which natural social processes in urban spaces are enhanced by data-driven protocols. Several research fields are already addressing this conceptual approach, identifying how public space analysis technologies can be employed to enable participatory urbanism (Cranshaw et al. 2012; Paulos et al. 2009). Foth (2018) describes the structured use of residents' data, feedback or input in planning as "participation in the "making of city" itself that re-conceptualises users as citizens (…) and residents as co-creators in a collaborative approach to citymaking and urban informatics" (Foth 2018, p. 10). The use of city user data in urban design decision-making processes offers a participatory approach that is scalable and which can operate as an ongoing process, transforming urban design practice from a linear and fixed outcome-oriented process into a dynamic management strategy.

5.4 Experiments in Generative Urban Design and Placemaking

The explore possibilities around the implementation of data-driven placemaking, a series of small experimental projects has been set up in the context of academic design + build workshops, which have been described in detail in our earlier publications (van Ameijde and Carlin 2012; van Ameijde et al. 2012; van Ameijde 2019). The projects explored workflow characteristics and conceptual processes in an incremental way, leading to a practical repertoire of hardware and software tools and processes, and to the refinement of our current research agenda, which focuses on the real-world implementation of data-driven placemaking methods.

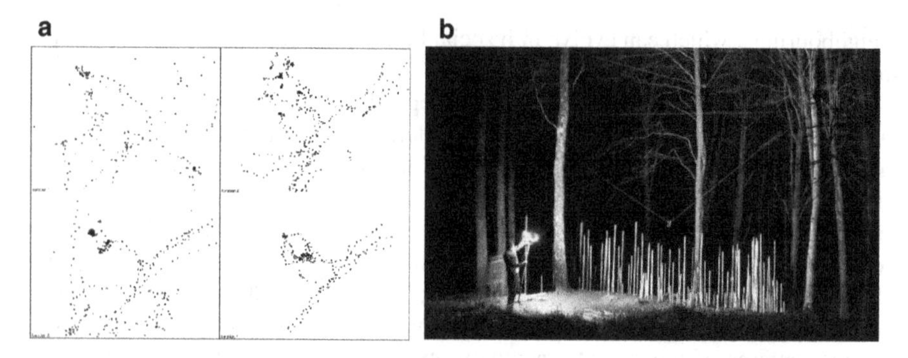

Fig. 5.3 **a** Iterative snapshots of location mapping of people, who are responding to the gradual refinement and densification around pathways and resting spaces. **b** Installation of the field of elements, using a webcam suspended from the trees and a cable robot as pointer device

The project titled 'Emergent Field' (Fig. 5.3a, b) explored a generative, rule-based design strategy that monitored people's movements at a forest site and materialised this as a field of timber stakes placed vertically within the terrain. The configuration was driven through a computational workflow, which through a webcam, suspended from the trees above the space, analysed people's locations and sent instructions to a cable robot system that guided human participants to install timber stakes around the perceived edges of circulation paths or resting spaces. The formation of the installation emerged throughout a series of iterations consisting of movement tracking, generative design translation and construction. The gradual refinement and articulation of circulation and inhabitation areas that occurred within both the digital design model and the physical space, allowed the final 'design' to be informed through the active negotiation between material and users around the real experience of the installation in the site.

A second project used the terrace of the Architectural Association as a testing ground, using a camera to record people inhabiting the terrace, documenting their position, duration of stay and distance to others (Fig. 5.4a, b). A set of computational rules was applied to the activity analysis, instructing human assistants to place furniture elements within the site. The experiments produced emergent outcomes, with an architectural structure that was grown over time without a predetermined design. Users interacted with the structure through sitting, leaning, placing coffee cups, etc., and generally staying longer and engaging in different activities than they would have normally done within this site. The experiments conducted as part of this project tested a rule-based urban growth scenario aimed at promoting social interaction, setting up an iterative scan and build process that incorporated feedback loops between behaviour mapping and construction.

A third project involved the creation of a pavilion, developed as a temporary installation at the central atrium of a large retail mall in Kuwait, aimed at creating three interconnected zones with increased privacy for socialising, rest or children's play activities (Fig. 5.5a, b). The pavilion helped to further develop our toolkit of

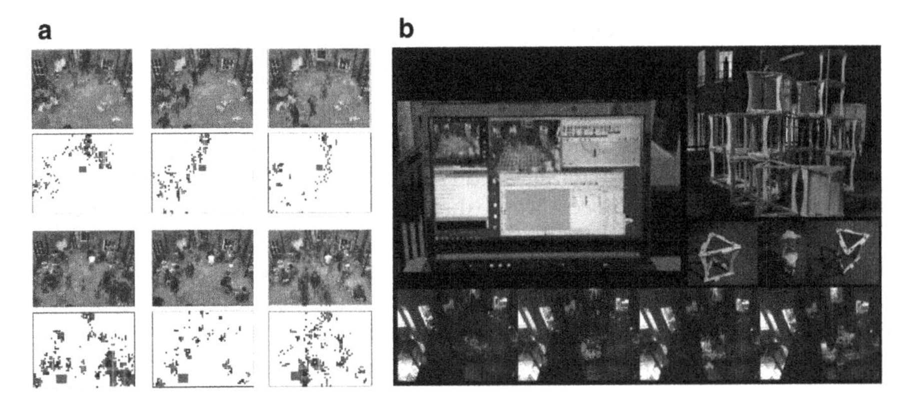

Fig. 5.4 **a** and **b** Snapshots of people monitoring as seen analysed with a webcam, informing the placement of furniture elements aggregating over time

computational tools and fabrication methods around the conceptual notion of user-driven generative design, using the scanning of people's movements and activities as a design driver.

The project was based on the conceptual framework of the SEEK project, incorporating an integrated information processing workflow consisting of (1) environmental scanning, (2) data analysis, (3) intervention decision and (4) environment mediation. These steps were implemented in the following manner:

(1) mapping of people's movements, locations and visibility fields from entrances;
(2) statistical analysis of visitor densities over time, mapped onto the space;
(3) a decision to locate three new activity spaces in strategic locations in relation to dense or sparsely visited spaces, and in relation to locations with high, medium or low visibility;

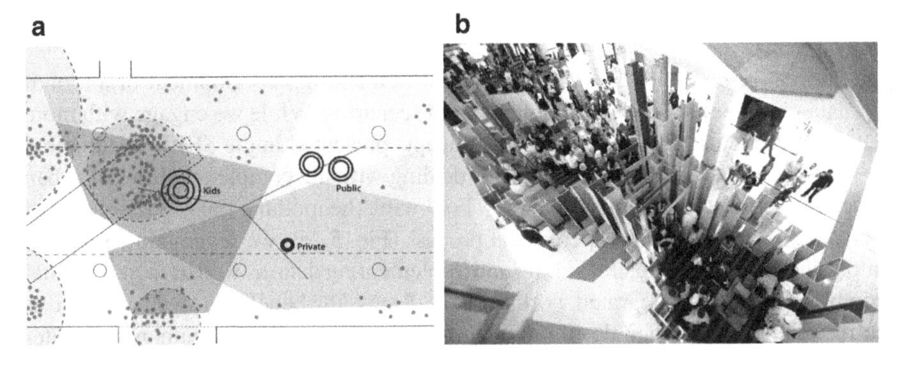

Fig. 5.5 **a** and **b** 'Emergent Constructions' pavilion design based on the mapping of visitor intensities and sight lines within a mall atrium

(4) a design system that generated a series of louvre elements of different densities and heights, to create boundary conditions that would provide various levels of isolation and enclosure.

While there were several different types of design decisions incorporated in this workflow, some relying on traditional forms of human preference and experience, the key idea embedded in the process was to directly connect each set of data throughout the different steps. By linking the computational analysis of the site to location and spatial quality decisions, a more informed and data-driven design proposal was produced. Connecting the desired spatial qualities of the new spaces to a generative design system that operated around similar spatial quality parameters, allowed to establish a meaningful feedback loop between local environmental analysis and the proposed intervention.

The 'Emergent Constructions' project explores two important dimensions crucial for the translation of Negroponte's ideas from the Lab to reality. Firstly, it incorporated a layered material system, that besides producing a building massing, delivered a range of other performative qualities. Secondly, the project showcased the notion of 'delayed' or 'indirect' participatory design, where user activities on site are used as a design driver. This mechanism was used to incorporate various technical aspects related to larger-scale architectural or public space construction projects. It also highlights the crucial question relating to the size and scope of data on which data-driven urban design decisions should be based. As direct participation or self-building scenarios might not necessarily reflect the consensus of a larger group of end-users, the recording of natural behaviours over a period of time might produce more accurate documentation on people's desired spatial activities and behavioural patterns.

5.5 Current Research into Data-Driven Urban Place Studies

Our current research focuses on applied research and the development of a toolkit for data-driven urban design implementation scenarios. While we engage with more detailed procedural translations of the concepts presented in the SEEK installation, we explore processes that are capable of dealing with the complex realities of urban spaces, their limitations and challenges. Following the modular conceptualisation of The Architecture Machine operational process (Fig. 5.2a), we separate the research into gathering, analysing, translating and implementing data, as well as their potential combination into an integrated workflow. We discuss the benefits, as well as the difficulties encountered when developing methods for more complex human–computer collaboration.

Environment Scanning

The first component of our workflow focuses on a scanning methodology capable of harvesting spatial data on the morphology of an urban site. For a reliable and detailed method of documenting site data, we employ unmanned aerial vehicle (UAV) based site scanning techniques using aerial photography and photogrammetry. Recent improvements in drone technology allow for the pre-programming of automated flight paths, camera stabilisation and tracking, which aids in the integration with other components of a workflow such as data processing, 2D environment mapping and 3D model reconstruction.

We employ photogrammetry software, which processes images captured with a UAV-mounted camera. A large number of photos is used to construct a three-dimensional model of the area of interest, which is then calibrated to match the dimensions and geolocation of the space in world coordinates. There is a wide range of applications such as the scanning of buildings and infrastructure for quality control, the mapping of landscape and site work, scanning of historic buildings for heritage conservation purposes, amongst others (Xu et al. 2013; Faltýnová et al. 2016; Golovina and Kanyukova 2016). UAVs can be pre-programmed to follow a specific trajectory that captures all angles and surfaces of the spaces and objects to be documented (Nex and Remondino 2013). The digital information is processed and combined in the photogrammetry software, which produces textured meshes based on different point cloud datasets produced by multiple drone photography missions.

Pedestrian Data Collection

Following our theoretical framework around the notion of placemaking for assessing the urban qualities of public spaces, we have developed an analytical methodology that is capable of documenting the social mechanisms within neighbourhoods, analysing the daily patterns of movements, activities and interactions. Extensive research into public spaces had been established beforehand by the sociologist William H. Whyte, who used mapping and direct observation techniques to deduce practical rulesets for creating successful public spaces (Whyte 1980). The study of human activities in public spaces promoted by Whyte and Jacobs has been further advocated by Gehl (1987, 2010, 2013). Whyte highlighted instinctive and cultural behaviours based on comfort and interpersonal relationships, while Gehl distinguished between 'necessary' and 'optional' activities to identify indicators of people's willingness to inhabit public space rather than pass through as quickly as possible.

Our workflow employs a mixed-method, qualitative and quantitative approach to the documentation of activities in public spaces, following the methodology for the ethnographic study of space outlined by Low (2000, 2016, 2019). The qualitative methodology combines several observational techniques aimed at forming a basic understanding of the patterns of behaviour in relation to context, including population counts, movement maps and behavioural maps (Low 2000). Following the findings of this qualitative study, a structured study can be conducted to collect quantitative data through observational studies of social behaviours. As part of our ongoing research, the use of digital cameras and image recognition software is being developed based on previously established methods for the analysis of people's movements and activities (Guo and Zhang 2020; Hanzl and Ledwon 2017).

In our current research, several UAV-mounted cameras are used to capture pedestrian locations in complex urban spaces, where due to obstructions and irregularities, a single camera would not be able to survey all of the spaces that combine into a continuous public space area. The videos are then processed and analyzed with Computer Vision Object Detection (CVOD) techniques to create digitalised pedestrian trajectory data. This data is integrated within the digital models of the urban space using perspective transform algorithms. Quantitative relationships between social activities and public space layout design are then extracted, spatialised and analysed in the Rhino/Grasshopper environment using customised Grasshopper definitions. Detailed technical descriptions of our workflow are presented in a separate recent publication (van Ameijde and Leung 2022).

Figure 5.6 shows the setup in one of our first experiments, in which four drones were used to obtain time-synchronised video footage from multiple aerial angles, to capture pedestrian locations during the evening rush hour. The drones were dispatched to different holding positions above the square at a flight level of 18 m above ground, covering different sections of the urban plaza. In separate tests, three 'dashcam'-type video cameras were installed in the same site, to overcome the problem of the limited flight time of the drones due to their battery life. Both approaches have different advantages and shortcomings, as the drones require minimal pre-planning and offer more freedom to capture strategic viewing angles. The building-mounted cameras offer the option to record over much longer timeframes, although their reduced image quality and lack of camera positioning data require adjustments in the analysis workflow. As a result, we envision employing both strategies in our future research, using drones for strategic snapshot observations of selected areas and using building-mounted cameras to investigate movement and activity patterns across larger timescales.

Pedestrian Tracking and Path Analysis

The video footage obtained through the site observations is further analysed using a combination of digital processes, including the most recently available applications of Machine Learning methodologies. Using the one-stage object detection model algorithm YOLOv4, our methodology uses a custom-trained dataset to recognise pedestrians from a bird's-eye viewing angle (Fig. 5.7). In the next step of data processing, the outcomes of the detection process are processed in DeepSORT (Simple Online and Realtime Tracking with a Deep Association Metric), which compares the changes

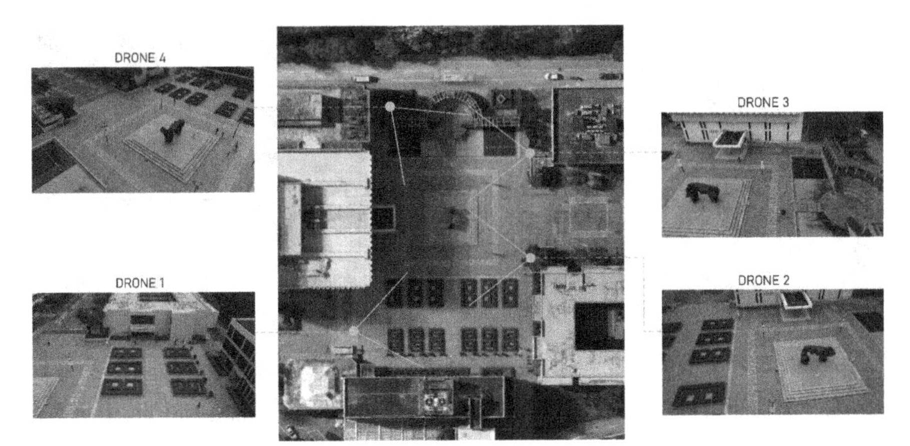

Fig. 5.6 Multiple UAV Deployment locations at the University Mall at the central campus of The Chinese University of Hong Kong

in moving elements in each current and previous frame, to make predictions about the past and future trajectory of that element (Wojke et al. 2017). This technique enables the continuous tracking of pedestrians even if their detection is lost in certain frames due to a missed detection or by the subject passing through covered areas.

After documenting data across specific periods of time, the movement and activity locations can be translated into time-based spatial data in the form of geolocation coordinate points. These can then be superimposed on the urban space environmental surface model which was calibrated against Open StreetMap's Geographic Information System (GIS). Both the manual mapping of activities by observers and the digital recording of people's locations in public spaces lead to the same outcome, a sequence of digital maps of people's location coordinates within the public space, organised in time intervals (Fig. 5.8).

Fig. 5.7 Object detection using Machine Learning methodologies

Fig. 5.8 Tracked trajectory data merged and projected on the photogrammetry model

People Location Analysis

To translate the basic data of movement and activity locations into meaningful insights about the intensities of use public spaces, basic statistical analysis can be performed through computational tools in Rhino/Grasshopper. The aim of this translation is to visualise the spatial use of public spaces in a compelling format, so that researchers, policymakers or local residents can engage in a conversation about the specific spaces and facilities that are successful, and which spaces are oversubscribed or underused. The key step in this process is a correlation analysis between recorded and geolocated coordinates of users, and the geolocated public space facilities such as seating, exercise equipment, canopies, playgrounds, etc. This correlation analysis can reveal which facilities are used more often, when and for how long, and how people move or interact around certain spaces.

A second analytical process to interpret people's location data is the analysis of *closeness*, the physical distance between people and "the in-between space that facilitates co-presence and regulates interpersonal relationships" (Madanipour 2003, p. 206). This distance indicates whether there might be occurrences of "co-presence" and "awareness", and a chance of social interaction (Hillier and Hanson 1984, p. 25). In the physical realm of public open spaces, we can evaluate various distances between actors based on the *proxemic interactions theory*, as defined by the anthropologist Edward T. Hall. Hall conceptualised personal space as a form of non-verbal and implicit communication, which he also referred to as the "silent language". While he emphasised that there may be social and cultural differences depending on the situation or location, his theory describes how people "in general perceive, interpret, structure, and (often unconsciously) use the micro-space around them, and how this affects their interaction and communication with other nearby people" (Marquardt and Greenberg 2015, p. 33). The physical proximity of people in public spaces indicates the potential for the forming of a community, with a virtual sphere for "probabilistic encounters" as a group of people may not have actually interacted yet, but are aware of each other's presence (Hillier and Hanson 1984; Hillier 1996; Major et al. 1997).

In our workflow, we employ a custom-built scripting tool to analyse the closeness of individuals using the discrete proxemic zones defined by Hall: *intimate* (0–0.5 m), *personal* (0.5–1 m), *social* (1–4 m) and *public* (>4 m) (Hall 1966). These values are supported by various scholars of personal space and interpersonal relationships, such as Hediger (1950), Sommer (1959), Altman (1975), and Bechtel and Churchman (2002). Our script analyses the distance to all other people within the public space, and groups and counts people who are within the thresholds of social and personal space. For visualisation purposes, a colour coding is added to the location points to indicate private individuals or couples with lighter colours, and social groups of three or more people with increasingly dark colours. Figure 5.9 illustrates this analysis, using one of the weekday snapshot observations produced for a research project focused on a Hong Kong public housing estate, mapped on the 3D model of the estate produced through photogrammetry.

In the final step of data translation, the human-centric analysis of people densities is translated into a space-centric analysis, defining the statistical occurrence of user presence and co-presence as a feature of the various locations within the case study space. The mapping process follows a basic logic of defining a spatial grid of cells,

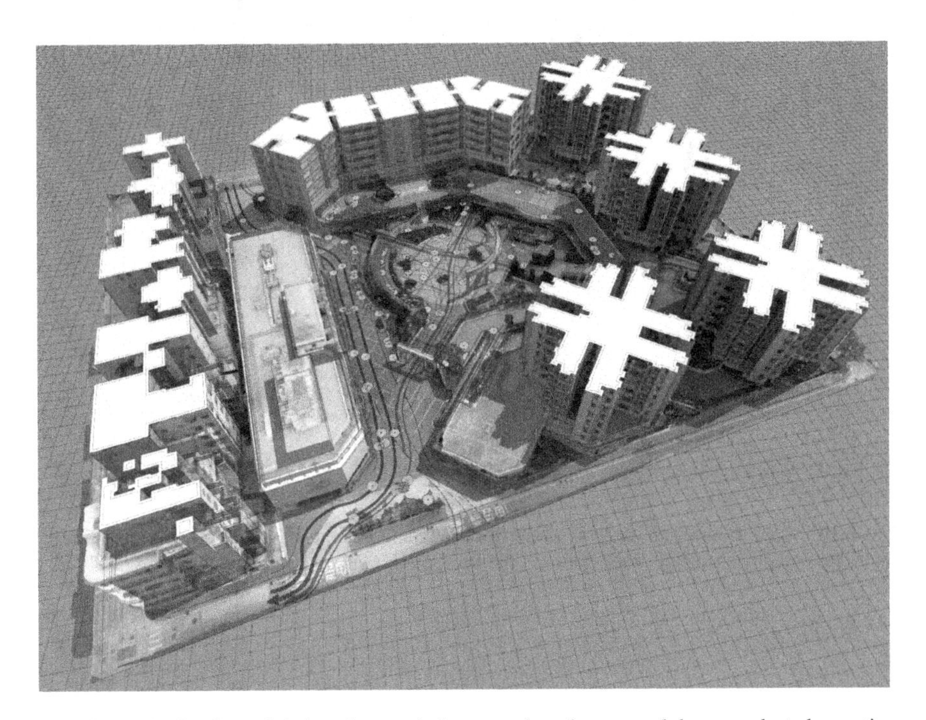

Fig. 5.9 Analysis of people's locations and closeness, based on a weekday snapshot observation at the Prosperous Garden housing estate. The people's location points are mapped on a truncated 3D model of the estate produced through photogrammetry

defined by their size (in the example shown, the grid spacing is set to two meters). The number of people's location markers is counted within each cell, and the social proximity value is also recorded. For this analysis, multiple datasets relating to various time intervals can be combined, produce insights into the general statistical patterns of space occupancy as they occur over longer periods of time. Figure 5.10 illustrates a data mapping of various snapshots of user locations across one typical weekday, compiled into one analytical visualisation projected onto the photogrammetry model.

As this paper focuses on a conceptual and procedural overview of the separate components of a data-driven urban design process, the detailed findings of the particular case study research shown here will be discussed in a different article. Instead, it is important to reflect on the critical overview and integration of the different stages of such a process, and how decisions around data interpretation can be guided by socially oriented policies.

Fig. 5.10 Space-centric analysis of people locations and closeness, based on a compilation of multiple weekday snapshot observations at the Prosperous Garden housing estate

5.6 Discussion

Data-Driven Design Policies

In this chapter, we have outlined separate but interrelated steps for the scanning and processing of environmental data, showcasing examples of public space mapping, urban morphology analysis and people location mapping and analysis. Following our interpretation of Negroponte's Architecture Machine concept as demonstrated in the 'SEEK' experiment, the data-driven nature of the public space analysis now allows us to conceive a design intervention in the same space, as part of a cybernetic system aimed at empowering its inhabitants. Negroponte's vision involves the setting up of a feedback loop between the analysed space, and a system of interventions which impacts the same parameters that were monitored in the first place. In our application of data-driven urban design, this implies an intervention protocol that could change, reduce or add public space elements that attract people to come to the area and stimulate private or social activities. The challenge in the translation of user and activity-related data toward design decisions lies in the formulation of policies for the evaluation of the data, and measures on how to respond to these findings.

As suggested in Fig. 5.2a, a series of Conditional Rules might be formulated, in relation to a Design Solutions Toolkit. These types of protocols are in essence not dissimilar from the public space management policies used in cities across the world today. Negroponte's vision to "eliminate the middleman" between the individual's needs and the incorporation of these needs into the built environment, seems in the context of urban design to align itself with the concept of 'the right to the city', which argues for "the right to belong to, and the right to co-produce the urban spaces" (Aalbers and Gibb 2014, p. 208). In the realm of urban design, honouring the concept of 'the right to the city' implies that we "emphasize the importance of the *use value* of urban space" rather than letting public space design and management be controlled by the interests of (adjacent) property owners (Purcell 2014, p. 142). Our interpretation of Negroponte's vision for the Architecture Machine is that data-driven urban design should not only facilitate the participation of ordinary citizens in the decision-making processes, but as a result of this setup—this should produce more diverse and inclusive public spaces that reduce social inequalities and segregation.

To achieve this goal, we should set up policies for public space design and management that aim to fulfil as many user requests as possible, with a minimal amount of coordination and regulation to resolve conflicting demands of different groups of end-users. This is in contrast with how various cities operate public space planning policies currently, as they often focus on minimising disturbance, accidents or maintenance costs by only facilitating activities and behaviours of a certain desirable range.

Data-Driven Design Systems

In our case study experiment at the Prosperous Garden estate, we speculate on a design strategy that matches supply to demand, providing public space facilities to the real-world usages as monitored. A permanent site monitoring system would be installed using CCTV camera feeds, using anonymized data collection as a formalised and structured commitment to performance evaluation (Fig. 5.11). If existing facilities would show to be unused, their number or location priority would be reduced, or they could be removed all together. If existing facilities would often be used to full capacity, additional numbers of those elements would be installed. An open-ended policy would be put in place to recognise and value activities with a wide range of characteristics. Specific design decisions for location choices would be based on statistical data, analysing which types of facilities in combination with what types of urban morphology characteristics have produced more frequent user engagement.

One of the key qualities of this implementation scenario is the deliberate lack of 'design' in the traditional sense, removing the notion of the 'masterplan' as an overarching, top-down framework of design decisions determined by planners. Instead, the project would make a catalogue of urban design elements available, able to be deployed onto the site in a range of configurations. Decisions on the implementation of changes on the site would be made autonomously by a system that continuously monitors human activities and interactions, guided by a set of policies that facilitate democratised and participatory modes of urban design, to produce a community-driven and supportive environment.

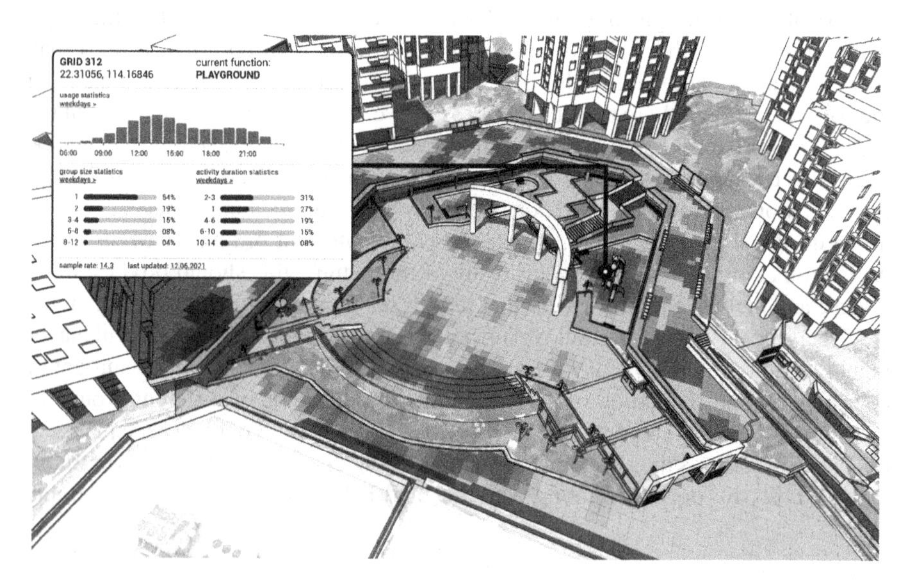

Fig. 5.11 User data evaluation on specific public space elements or locations

Critical and Ethical Considerations Around Data-Driven Urban Design

While the practice of collecting data on urban spaces and city users has been in use since well before the emergence of ubiquitous computing (Fitzgerald 2016; Miller 2018), the scale and automation of the smart city monitoring methods signal the potential for urban governance to transform from being informed by data, to being data-driven (Kitchin 2017). The implied autonomous nature of these mechanisms, their universal distribution and potential lack of transparency raise important questions about the forms of governance enabled by these technologies (Cardullo 2020).

Several smart city projects that included video analytics have been the subject of citizens' concerns (Mann et al. 2020; O'Malley and Smith 2020), even as their operating protocols had been set up to only collect anonymous and statistical data. Examples from around the world show how the introduction of systems such as high-definition CCTV cameras and smart lampposts can be associated with increased government control (Crawford 2014) or the commodification of public space user data through 'surveillance capitalism' (Zuboff 2019). Ratti and Claudel (2016) advocate for the use of localised, non-profit and community-driven projects and use project websites and open data protocols that enhance transparency and public accountability. Cardullo (2020) responds to common concerns around smart city systems by identifying strategies that build trust and social capital, through the explicit use of ethical data management policies, based on civil, social and digital rights.

Instead of seeing citizens as data providers or as people that need to be monitored, nudged or controlled (Cardullo and Kitchin 2019), the workflow presented in this chapter aims to enable a genuine human-centric and citizen empowering participatory process. As Lefebvre's 'right to the city' concept implies citizen's right to 'co-creation', it seeks to engage local communities in discussion, negotiation, criticism and the proposing and implementation of alternatives based on their experiences with the limitations and opportunities of their public spaces. More research is needed to explore how the data monitoring, analysis and intervention strategies are made accessible, legible and debatable through transparent and easy to understand interfaces and communication protocols, to enable these systems to earn the trust of the participating public.

5.7 Conclusions

In this article, we have traced some of the ambitions of early explorers of cybernetic systems in architecture and urbanism, and attempted to reposition these in the context of the emerging opportunities of public space design in the age of ubiquitous computing. The visionary scenarios of Negroponte, Pask and Friedman highlighted issues with the 'ego' and limitations of a single human designer and proposed that the objective and systematic nature of machinic systems of control would be better suited to translate the complex and collective requirements of a group of end-users into appropriate design solutions. As identified by Negroponte and emphasised in this article, autonomous systems could produce desirable or highly undesirable outcomes, and 'architecture machines' would have to be sensitive and open to humanistic criteria and interventions. The current debate about smart city applications conceived within neoliberal cultures of government highlights the need for further research into how these initiatives can be set up to serve the public good.

The key issue is to focus on the overarching objectives governing these systems, to verify whether data-driven systems produce meaningful management scenarios that implement positive social policies in relation to human-centric urban policies, and in line with fundamental principles such as the notion of 'the right to the city' and that of 'spatial justice'. The real-world implications of these theoretical concepts involve social, political and economic rights and the production of urban spaces that contributes to developing people rather than excluding or exploiting them.

In our own data-driven design experiments and speculations, our strategies focus on facilitating the pluralistic demands for facilities in housing estate public spaces, as these serve a multitude of purposes including facilitating active and passive recreation, socialising, community formation and participation. In our case study of housing estate, there are user groups from different age groups, income levels, ethnicities and cultures. An open-ended approach to data gathering and analysis would be able to learn about the requirements and desires of each of these groups, without making specific presumptions or prescriptions.

The central component of Negroponte's 'architecture machine' is the ideology of the feedback loop—setting up cybernetic systems that use data on the successful or unsuccessful deployment of urban design elements for the continuous evaluation and updating of the built environment. These new types of computational systems and emerging modes of practice allow us to start testing real-world scenarios of intelligent adaptive environments. These opportunities demand a rethinking of the roles and responsibilities of urban designers, as smart city critics urge for the calibration of data-driven design policies according to a societal vision that promotes citizenship, social justice and the right to the city. As the field of urban informatics opens up new ways of translating human behaviour and cultural and social interactions into urban design outcomes, there is an urgent need to consider how data management mechanisms can inclusively improve the quality of life in future urban communities.

Acknowledgements The 'Emergent Field' project was developed by Jeroen van Ameijde in collaboration with Jenny Hill and Dennis Vlieghe. The 'Public Space Furniture' project was developed by Christine Cai, Sungbin Ryoo and Ke Wang, taught by Jeroen van Ameijde and Brendon Carlin. 'Emergent Constructions' was developed by Jeroen van Ameijde, Sulaiman Alothman, Mohammed Makki, Yutao Song and Manja van de Worp. The ongoing research into UAV-based people location tracking and analysis is being conducted by Jeroen van Ameijde and Carson, Ka Shut Leung. We thank the two anonymous peer reviewers whose generous comments have contributed to the improvement of this work.

References

Aalbers MB, Gibb K (2014) Housing and the right to the city: introduction to the special issue. Int J Hous Policy 14(3):207–213

Altman I (1975) The environment and social behavior: privacy, personal space, territory, and crowding. Brooks/Cole Publishing Company, Monterey, California

Barns S (2020) Platform urbanism: negotiating platform ecosystems in connected cities. Springer. https://doi.org/10.1007/978-981-32-9725-8

Bechtel RB, Churchman A (2002) Handbook of environmental psychology. Wiley

Breuer J, Pierson J (2021) The right to the city and data protection for developing citizen-centric digital cities. Inf Commun Soc 24(6):797–812. https://doi.org/10.1080/1369118X.2021.1909095

Cardullo P (2020) Citizens in the 'smart city': participation, co-production, governance. Routledge

Cardullo P, Kitchin R (2019) Being a 'citizen' in the smart city: up and down the scaffold of smart citizen participation in Dublin, Ireland. Geojournal 84(1):1–13

Cranshaw J, Schwartz R, Hong J, Sadeh N (2012) The livehoods project: utilizing social media to understand the dynamics of a city. In: Proceedings of the international AAAI conference on web and social media, vol 6, no 1, pp 58–65

Crawford K (2014) The anxieties of big data. New Inq 30:2014

Cresswell T (2014) Place: an introduction. Wiley

De Lange M (2019) The right to the datafied city: interfacing the urban data commons. In: The right to the smart city. Emerald Publishing Limited, pp 71–83

Deleuze G, Guattari F (1987) A thousand plateaus: capitalism and schizophrenia. University of Minnesota Press, Minneapolis

Dyer M, Corsini F, Certomà C (2017) Making urban design a public participatory goal: toward evidence-based urbanism. Proc Inst Civ Eng-Urban Des Plan 170(4):173–186

Faltýnová M, Matoušková E, Šedina J, Pavelka K (2016) Building facade documentation using laser scanning and photogrammetry and data implementation into BIM. Int Arch Photogramm, Remote Sens Spat Inf Sci XLI-B3:215–220

Fitzgerald M (2016) Data-driven city management: a close look at Amsterdam's Smart City initiative. In: MIT Sloan Management Review. https://sloanreview.mit.edu/case-study/data-driven-city-management/

Foth M (2017) Lessons from urban guerrilla placemaking for smart city commons. In: Proceedings of the 8th international conference on communities and technologies. ACM, pp 32–35. https://doi.org/10.1145/3083671.3083707

Foth M (2018) Participatory urban informatics: towards citizen-ability. Smart Sustain Built Environ 7(1):4–19. https://doi.org/10.1108/SASBE-10-2017-0051

Foth M, Brynskov M, Ojala T (eds) (2016) Citizen's right to the digital city: urban interfaces, activism, and placemaking. Springer. https://doi.org/10.1007/978-981-287-919-6

Foth M, Caldwell G, Fredericks J, Volz K (2018) Augmenting cities beyond bedazzlement: empowering local communities through immersive urban technologies. In: Workshop proceedings of augmenting cities and architecture with immersive technologies, Media Architecture Biennale (MAB-18). Media Architecture Biennale, pp 1–4

Friedmann J (2010) Place and place-making in cities: a global perspective. Plan Theory Pract 11(2):149–165

Gehl J (1987) Life between buildings: using public space. Van Nostrand Reinhold, New York

Gehl J (2010) Cities for people. Island Press, Washington, DC

Gehl J, Svarre B (2013) How to study public life. Island Press/Center for Resource Economics, Washington, DC

Guo, Zhang (2020) Observations of urban activities with computer vision. In: Proceedings of the 13th international forum on urbanism (IFoU) online congress, urbanism in the mobile Internet era, Nanjing University, China, 9–12 Oct 2020

Hägerstrand T (1953) Innovationsförloppet ur korologisk synpunkt, vol 25. Gleerupska univ.-bokhandeln

Hall ET (1966) The hidden dimension, 1st edn. Doubleday, Garden City, NY
Hanzl M, Ledwon S (2017) Analyses of human behaviour in public spaces. In: Wang H, Ledwon S (eds) Proceedings of joint conference ISOCARP-OAPA - 53rd ISOCARP congress, smart communities, Portland, Oregon, USA, 24–27 Oct 2017
Hediger H (1950) Wild animals in captivity. Butterworths Scientific Publications
Hillier B (1998) Space is the machine: a configurational theory of architecture. Cambridge University Press, Cambridge
Hillier B, Hanson J (1984) The social logic of space. Cambridge University Press, Cambridge
Jacobs J (1961) The death and life of great American cities. Vintage Books, New York
Kitchin R (2015) Making sense of smart cities: addressing present shortcomings. Camb J Reg Econ Soc 8(1):131–136
Kitchin R (2017) Data-driven urbanism. In: Kitchin R, Lauriault T, McArdle G (eds) Data and the city. Routledge, London, pp 44–56. https://doi.org/10.4324/9781315407388-4
Koolhaas R (2014) Rem Koolhaas asks: are smart cities condemned to be stupid? ArchDaily, 10 Dec 2014
Lau KY, Murie A (2017) Residualisation and resilience: public housing in Hong Kong. Hous Stud 32(3):271–295
Lee S, Bier H (2019) Apparatisation in & of architecture. SPOOL 6(1):3–4
Lefebvre H (1968) Le droit a la ville. Anthropos, Paris
Legeby A (2013) Patterns of co-presence: spatial configuration and social segregation. Doctoral dissertation, KTH Royal Institute of Technology
Leszczynski A (2015) Spatial big data and anxieties of control. Environ Plan D, Soc Space 33(6):965–984. https://doi.org/10.1177/0263775815595814
Low S (2000) On the plaza: the politics of public space and culture. University of Texas Press, Austin, TX
Low S (2016) Spatializing culture: the ethnography of space and place. Routledge, London, England; New York, New York
Low S, Simpson T, Scheld S (2019) Toolkit for the Ethnographic Study of Space (TESS). Public Space Research Group, Center for Human Environments, The Graduate Center, City University of New York
Madanipour A (2003) Social exclusion and space. In: LeGates R, Stout F (eds) The city reader, 3rd edn. Routledge, London, pp 181–189
Major MD, Stonor T, Penn A, Hillier B (1997) Housing design and the virtual community. In: Children and youth in the city. 19th international making cities livable conference. Charleston, South Carolina
Marquardt N, Greenberg S (2015) Proxemic interactions, from theory to practice. Synthesis lectures on human-centered informatics. Morgan & Claypool, San Rafael
Miller SR (2018) Urban data and the platform city. In: Davidson N, Finck M, Infranca J (eds) Cambridge handbook of law and regulation of the sharing economy. Cambridge University Press, Cambridge, UK, pp 192–202
Negroponte N (1970) The architecture machine. MIT Press, Cambridge, Mass
Negroponte N (1975) Soft architecture machines. MIT Press, Cambridge, Mass.
Nex F, Remondino F (2013) UAV for 3D mapping applications: a review. Appl Geomat 1–15. https://doi.org/10.1007/s12518-013-0120-x
Nisha B, Nelson M (2012) Making a case for evidence-informed decision making for participatory urban design. URBAN DES Int 17:336–348
O'Malley P, Smith GJD (2020) "Smart" crime prevention? Digitization and racialized crime control in a smart city. Theor Criminol 1362480620972703. https://doi.org/10.1177/1362480620972703
Pask G (1969) The architectural relevance of cybernetics. Archit Des 39(9):494–496
Paulos E, Honicky RJ, Hooker B (2009) Citizen science: enabling participatory urbanism. In: Handbook of research on urban informatics: the practice and promise of the real-time city. IGI Global, pp 414–436
Pavlovic M, Bier H, Pillan M (2020) Ambient UX for cyber-physical spaces. SPOOL 7(3):27–36

Pierson J (2012) Online privacy in social media: a conceptual exploration of empowerment and vulnerability. Commun Strateg 88:99–120
Pillan M, Pavlovic M, Bier H (2020) Towards an architecture operating as a bio-cyber-physical system. SPOOL 7(3):47–58
Purcell M (2014) Possible worlds: Henri Lefebvre and the right to the city. J Urban Aff 36(1):141–154
Ratti C, Claudel M (2016) The city of tomorrow: sensors, networks, hackers, and the future of urban life. Yale University Press, New Haven; London
Rowe J (1972) Life in a computerised environment. In: Electronics Australia, September 1972
Samuelsson K, Barthel S, Colding J, Macassa G, Giusti M (2020) Urban nature as a source of resilience during social distancing amidst the coronavirus pandemic, no 3wx5a. Center for Open Science
Shaw J, Graham M (2017) An informational right to the city? Code, content, control and the urbanization of information. Antipode 49(4):907–927
Shi W, Goodchild M, Batty M, Kwan MP, Zhang A (2021) Urban informatics. Springer, Singapore. https://doi.org/10.1007/978-981-15-8983-6. ISBN 978-981-15-8982-9
Sommer R (1959) Studies in personal space. Sociometry 22:247–260. https://doi.org/10.2307/2785668
Steenson M (2017) Architectural intelligence: how designers and architects created the digital landscape. The MIT Press, Cambridge, Massachusetts
Stokols D (1990) Instrumental and spiritual views of people-environment relations. Am Psychol 45(5):641
Talen E (1999) Sense of community and neighbourhood form: An assessment of the social doctrine of new urbanism. Urban Stud 36:1361–1379. https://doi.org/10.1080/0042098993033
Tombs S, Whyte D (2015) Counterblast: crime, harm and the state-corporate nexus. Howard J Crim Justice 54(1):91–95
Tuan Y (1976) Humanistic geography. Ann Assoc Am Geogr 66(2):266–276
van Ameijde J (2019) The architecture machine revisited, experiments exploring computational design and build strategies based on participation. In: Lee S, Bier H, TU Delft (eds) Cyber-physical architecture #2. SPOOL
van Ameijde J, Carlin B (2012) Digital construction: automated design and construction experiments using customised on-site digital devices. In: Achten H, Pavlicek J, Hulin J, Matejovska D (eds) Digital physicality - proceedings of the 30th eCAADe conference, vol 2. Czech Technical University in Prague, Prague, pp 439–446
van Ameijde J, Leung KS (2022) UAV-based people location tracking and analysis, for the data-driven assessment of social activities in public spaces. In: 27th international conference on computer-aided architectural design research in Asia, CAADRIA 2022. The Association for Computer-Aided Architectural Design Research in Asia (CAADRIA)
van Ameijde J, Carlin B, Vlieghe D (2012) Emergent constructions: experiments towards generative on-site design and build strategies using customised digital devices. In: ACADIA 12: synthetic digital ecologies - proceedings of the 32nd annual conference of the association for computer aided design in architecture (ACADIA). California College of the Arts, San Francisco, pp 539–545
van Asselt MBA, Rijkens-Klomp N (2002) A look in the mirror: reflection on participation in integrated assessment from a methodological perspective. Glob Environ Chang 12(3):167–184
van Dijck J (2014) Datafication, dataism and dataveillance: Big Data between scientific paradigm and ideology. Surveill Soc 12(2):197–208
van Doorn N (2018) The parameters of platform capitalism. Krisis: J Contemp Philos 1:103–107. https://krisis.eu/the-parameters-of-platform-capitalism/
Vanolo A (2014) Smartmentality: the smart city as disciplinary strategy. Urban Stud 51(5):883–898
Whyte WH (1980) The social life of small urban spaces. Conservation Foundation, Washington, D.C.
Wojke N, Bewley A, Paulus D (2017) Simple online and realtime tracking with a deep association metric. In: 2017 IEEE international conference on image processing (ICIP). IEEE, pp 3645–3649

Xu S, Vosselman G, Oude Elberink S (2013) Detection and classification of changes in buildings from airborne laser scanning data. ISPRS Ann Photogramm, Remote Sens Spat Inf Sci II-5-W2(5):343–348

Yin L, Shaw SL (2015) Exploring space–time paths in physical and social closeness spaces: a space–time GIS approach. Int J Geogr Inf Sci 29(5):742–761

Zuboff S (2019) The age of surveillance capitalism: the fight for a human future at the new frontier of power: Barack Obama's books of 2019. Profile Books

Part II
Architectural Intelligence, Machine and Human Learning

Chapter 6
Architectural Intelligence, Machine, and Human Learning

Philippe Morel

While it was referred to as postmodern (Lyotard 1979), from the 1960s until around the end of the 1990s, the information society was still often treated as a modern society in which the service sector would have taken precedence over the manufacturing and material goods circulating according to a new logic of networks (Castells 1996–1998) which also governed human relations, relations of power or simple relations of friendship. In line with its original name given by its creator Claude Shannon (Shannon 1948)—a (mathematical) "theory of communication"—this information society was also named the communication society. In architecture, this communicational aspect was confirmed by Robert Venturi who affirmed that "*modern architecture is about space, postmodern architecture is about communication*" (Venturi 1996). Although no one can deny the current importance of the communication phenomenon, we would nevertheless be wrong to limit ourselves to it. Our world, in fact, is at the same time *informational, communicational,* and *computational* and it is indeed this triple nature that is an urgent question today. Matter has not disappeared; it is as crucial as in all previous eras, but it is nowadays dominated by information. It is either a source of "raw informational material," or a vector of information, or both. It is also a support of computation that can be programmed as desired according to the most diverse models of computation. What are the consequences of these transformations, or rather, among the massive consequences of these transformations and within the architectural discipline, which ones will be discussed in this section dedicated to architectural intelligence, machine, and human learning?

The first of four chapters, by Dr. Roberto Bottazzi (The Bartlett School of Architecture-University College London), entitled *Architectural Knowledge and Learning Algorithms*, will concern the new conditions that govern architectural knowledge in the age of machine learning algorithms. The interactions between

P. Morel (✉)
The Bartlett School of Architecture, University College London, London, UK
e-mail: p.morel@ucl.ac.uk

P. Morel and H. Bier (eds.), *Disruptive Technologies: The Convergence of New Paradigms in Architecture*, Springer Series in Adaptive Environments,
https://doi.org/10.1007/978-3-031-14160-7_6

humans and the ever more complex and foreign field of these algorithms will be explored, in relation to the "*complexity and cultural richness incorporated in the thought automation project*". As R. Bottazzi notes, we must go beyond the mere understanding of the technical (algorithmic) aspect of the learning problem alone since "*learning algorithms pose more complex and conceptual challenges as they suggest a radical reorganization of space and scale.*" These algorithms reorganize not only space but also the representation of urban complexity provided by data, while by organizing these data they, in turn, provide us with new representations, certainly intelligible, but partial.

The second chapter—*On Legibility: Machine Readable Architecture*—by Associate Professor Andrew Witt (Harvard Graduate School of Design), deals with the concept of architectural and computational readability, encoding, and visual language in architecture. As one will see, A. Witt proposes "*three related frames through which to interpret the entangled practices of architectural and machine readability. The first is a capsule chronology of machine readability, from its roots in tabular statistical datasets in the nineteenth century to its convergence with AI and machine learning today [...]. The second is an examination of the concept of architectural readability as it evolved complimentarily in the 1970s [...]. The third explores the intersections of the first two through the presentation of two design projects that use machine vision and machine readability [...].*" According to A. Witt, new ways of reading and generating architectural forms are needed, through the concept of machinic reading. "*From projects that morphologically catalog the world's billion buildings to the application of shape classification for radical waste reuse,*" this "*machinic reading is transforming the roles and products of design.*"

The third chapter, in this section *Architectural Intelligence, Machine, and Human Learning*, by Dr. Theodore Spyropoulos (Architectural Association School of Architecture), is entitled *Where is reality? Can you show it to me? Constructing Artificial Agency.* It goes back to one of the founding theories of the information society, i.e., cybernetics: the first cybernetics but also the one called "second cybernetics" (or "higher order cybernetics"). Summoning the English psychiatrist William Ross Ashby in his book titled *An Introduction to Cybernetics* (1956), Th. Spyropoulos insists on the fact that while cybernetics is a "*theory of machines*" it deals not with objects but with "*ways of behavior*". "*It does not ask "what is this thing?" but "what does it do?"*". Following in this the reading by Johnston (2008) of W. R. Ashby's cybernetics, Th. Spyropoulos notes that the real object of this theory is the "*domain of all possible machines.*" That some of these machines were not made by Man or by Nature is a secondary question. What cybernetics truly offers is a "*framework on which all individual machines may be ordered, related and understood.*" For Th. Spyropoulos, what matters is not machines as such, as informational machines, but their behavior and communication potential within this framework common to machines and humans. Starting from questions about (massive) communication, Th. Spyropoulos also poses the ontological question of the nature of reality. Where is she? In the minds of humans or in the memories of machines?

Following these three chapters, the last one—*From Disruptions in Architectural Pedagogy to Disruptive Pedagogies for Architecture*—, by Dr. Sevgi Türkkan

(Istanbul Technical University), concerns, a fortiori in view of the radical transformations that we have just mentioned, the key question of education. This chapter is a pedagogically oriented paper aiming at the techno-cultural-pedagogic shift in architectural education. A manifesto for forms of intelligence, labor, creativity, and reorganization of space offered for more relevant architectural learning. It is calling for radical changes in the pedagogic agenda thanks to recent advancements in digital knowledge, big data availability, and open-source AI tools. It challenges in a very concrete manner the "*mainstream and "ordinary" architecture school, its educational concepts, curriculum, pedagogic rituals, values, and the disciplinary ethos that lies underneath.*" As S. Türkkan mentions, the aim of this chapter is "*to outline trajectories for this agenda, by raising a series of questions regarding architectural learning and the role of institutions in the twenty-first century*".

References

Castells M (1996–1998) The rise of the network society, the information age: economy, society and culture, vol I. Blackwell, Malden, MA; Oxford, UK; The power of identity, the information age: economy, society and culture, vol II. Blackwell, Malden, MA; Oxford, UK; End of millennium, the information age: economy, society and culture, vol III. Blackwell, Malden, MA; Oxford, UK
Johnston J (2008) The allure of machinic life: cybernetics, artificial life, and the new AI. The MIT Press
Lyotard J-F (1979) La Condition postmoderne. Les éditions de minuit, Paris
Shannon CE (1948) A mathematical theory of communication. Bell Syst Tech J 27(3):379–423
Venturi R (1996) Iconography and electronics upon a generic architecture: a view from the drafting room. MIT Press

Chapter 7
Architectural Knowledge and Learning Algorithms

Roberto Bottazzi

7.1 Introduction

All architecture is now designed with the help of computers marking one of the most fundamental changes in the history of the discipline. The digitisation of architecture has not been a smooth, linear process as computers have also offered a new conceptual lens through which to critically question the tenets of the profession and its practices. The transformations triggered by digital technologies are far from over as the simultaneous increases in data availability, computational power and effectiveness of automated processes such as machine learning (ML) methods continue agitating the discipline of architecture and affect its future.

The promise of a more efficient, streamlined, 'unproblematic' design process often accompanies the introduction of new digital tools in design. This paper will task to challenge this view to demonstrate that the penetration of new techniques in the history of architecture has not happened under the auspices of efficiency only, but rather through a cultural ambition to introduce new ways of thinking about space and design. Images of efficiency accompanying the introduction of digital tools in design practice are however more rhetorical than actual as they clash against the complexity of effectively learning and deploying advanced modellers or new programming languages; a process that requires time, dedication, and, most importantly, a knowledge of many different disciplines. To master a certain computational technique, one has to greatly exceed the acquisition of mere technical skills. In the process of learning a particular digital skill, one becomes increasingly aware of the different complex ideas underpinning digital techniques and that their application to design is far more profound and exciting than the prospect of optimising the status quo.

R. Bottazzi (✉)
The Bartlett, UCL, London, UK
e-mail: roberto.bottazzi@ucl.ac.uk

P. Morel and H. Bier (eds.), *Disruptive Technologies: The Convergence of New Paradigms in Architecture*, Springer Series in Adaptive Environments,
https://doi.org/10.1007/978-3-031-14160-7_7

The introduction of learning algorithm in design marks a discontinuity with traditional technologies of design as well as with the digital techniques which characterised the 'first digital turn' in the 1990s (Carpo 2017). The root of such discontinuity is represented by data both understood quantitatively (that is, the massive increase in the ability to sense and store ever larger datasets) and qualitatively (represented by the recent improvement of algorithms to access, structure and manipulate enormous datasets). The combination of large datasets and instruments to interact with them radically changes how space is represented and designed. We can speak of radical transformations because the ML methods confront architects with a new representation of space: for instance, by operating at unprecedently large scales while maintaining a very high level of resolution, or by comparing and combining data on very different aspects of space with the result of collapsing disciplinary boundaries. Hayles (2014) speaks of computational machines as 'nonconscious'; that is, as something altogether different from human cognitive mechanisms and affordances. The task, in fact, is not to restore some sort of previous condition to re-establish the primacy of human thinking. Rather, the ambition is to exploit the gap between the different modes of apprehension of objects displayed by humans and the alien qualities of Artificial Intelligence (Parisi 2019). The challenge for spatial designers is to establish models for communication with the 'alien' quality of computation. These considerations extend to how algorithms understand space as well; well-established architectural notions to categorise space such as scale, granularity, variation and compound ones such as site, programme, scale, type are all challenged by the introduction of ML methods.

This paper proposes to conceptualise the introduction of ML methods in design through the lens of cryptography; a discipline that has been historically central to computation. Cryptography, in fact, concentrates on the techniques of exchange between noisy, random, 'meaningless' domains and intelligible ones; a problem that is also central to the implementation of ML methods in architectural design. In the discussion that follows, cryptography is tasked to provide ways of thinking and instruments to navigate massive datasets and interfaces with the 'alien' operations of learning algorithms. The design process could be understood as an analogous to cryptography as it consists of devising methods to extend and improve the representation of space in order to manipulate it. New techniques in design often emerge when new affordances are required to deal with a mutated environment. For instance, the most important design technology, drawing, enters architecture at the end of the Gothic period when the complexity of building sites made drawings essential to design and build intricate structures. The 'alien' quality of ML algorithms calls for ways of thinking about the interaction between humans and machines in design.

To grasp how a cryptographic approach to design could be articulated, we will first define how traditional computational machines (analogue and code-based ones) represent and conceptualise real phenomena. Comparing analogue and code-based computing contraptions will help us foreground the epistemological value of design techniques and the challenges that the introduction of ML methods pose. The figure of the prosthesis will be a useful reference to accompany this journey as it will allow us to conceptualise how we explore and interact with unknown territories to gather knowledge. The second part of the chapter will focus on cryptography by investigating how it emerged as computational technique and, then, by thinking about what implications it can have on creative processes in design. The conceptual figure accompanying this part of the discussion will be that of Cypher: the very mechanism regulating the interaction between noisy and intelligible datasets.

7.2 Prosthetics

To frame the discussion on the relation between learning algorithms and design we will begin from the notion of prosthetics understood as a conceptual model that links techniques to the emergence of knowledge. Though the subject has been extensively investigated, our argument quickly retraces some of the key definitions to adapt them to the new issues presented by the introduction of learning algorithms in design. The term prosthesis derives from two Greek words: *Protíthemi* (to present, to expose, assign) and *Prostíthemi* (put next, add, stick to).[1] The first origin points at the relation between prosthetics and knowledge, whereas the second suggests that prosthetic objects are added or applied to other objects or contexts in order to activate them. As Sini (2009) observes,the mechanisms of prosthetic objects can already be observed at work in simple cases such as in the use of a walking stick by a blind person. We can speak of the stick as a prosthetic object because it is 'added' and not part of the body of the person holding it, whilst the stick prolongs the person's cognitive abilities which extend to include the prosthetic object (what philosophers call 'an extension of mind' (Clark and Chalmers 1998)). The stick 'exposes' what is otherwise inaccessible to the blind person's senses by foregrounding aspects of the environment that can now be learnt. The process is continuous as knowledge can never fully coincide with its object that is being investigated; if it did, the whole epistemological process would come to an end. The labour of the prosthetic instrument is to activate, to mobilise something unfamiliar or unknown in order to make it intelligible.[2]

The conceptual figure of the prosthesis is a useful analogy to begin to grasp the epistemological and design issues emerging from the application of ML methods to architecture. The multi-dimensional space of massive datasets is an uncharted, open territory to explore which algorithms can survey, mobilise knowledge embedded in data, and make it intelligible. Intelligibility is a property of design techniques; that is, design techniques can be considered to be prosthetic devices that can be applied to an unknown problem in order to make it intelligible and amenable to manipulations. We are still at a stage that precedes the crystallisation of signs into meaning; rather, prosthetic techniques provide design methods with an interface to manipulate noisy, unstructured domains. As we shall later discuss, the introduction of ML

[1] "Prosthesis". Oxford English Dictionary. Oxford: Oxford University Press.

[2] Nietzsche defined knowledge as the process of moving something from the domain of the unknown or stranger into that of the known or familiar (Nietzsche, cited in Sini 2009, p. 48).

methods in design greatly enhances the capacity to navigate the complex, 'meaningless' environment of data to the point that the notion of prosthetic may no longer be sufficient to account for the processes at work. Particularly, the figure of the prosthesis may struggle to account for two conditions related to the application of ML methods to design processes. First, learning algorithms simultaneously analyse data and produce new datasets themselves. ML methods conflate analysis and production, whereas prosthetic objects limit their action to surveying. Second, learning algorithms rely on statistical rather than deterministic mathematics thus setting up a dynamic, interactive relation between noisy and organised data. The complex algorithmic process of correlation and generation thus opens up a playful, dynamic space for speculative thinking in which different parameters and approaches can be tested to probe datasets. This point seems to push the notion of prosthetic to its limit and call for the introduction of other concepts and methodological figures to update a pure prosthetic reading of ML methods.

The issues outlined in this paragraph are not new to architectural discourse which has a long history of inventing computational machines in order to extend the capacity to represent, conceptualise and manipulate space. The next paragraph will compare the main characteristics of physical/analogue design machines and code-based/discrete ones to foreground how each technique impacts the representation of space and what affordances it provides (Fig. 7.1).

Fig. 7.1 N^2P^2—Neural Network Public Places. Project for the reorganisation of the Île de la Cité in Paris. Neural Network navigating the roads of Paris. Authors: R. Bottazzi, T. Varoudis, P. Prajapati, X. Wang

7.3 Architectural Computations: Perspective Machines and Code

Architects have always built machines. Vitruvius dedicated one of the books of his treatise to this very subject. After all, the limits of architects' imagination coincide with those of the instruments they use. Technologies format space according to precise criteria that determine how spatial knowledge emerges and, in turn, what agency such knowledge will have on the design process.

The first wave of machines systematically developed by architects and artists to support design coincided with the introduction of linear perspective in the Renaissance. The central aim of the development of such machines was the mechanisation of the theories of sight that underpinned the construction of mathematical perspective. Such contraptions represented an attempt to achieve three distinct objectives: to reify human faculties in order to extend them, expand artists' and architects' possibilities for representation, and, finally, automate bodily gestures. Their purpose was hardly to make the creative process more efficient as we have little evidence that the designs depicted in drawings and etchings were built, or utilised (Kemp 1990, pp. 167–220).[3] Perspective machines, however, offer a series of valid insights on the relationship between automated thinking and design that can still shine some light on present issues accompanying the introduction of learning algorithms in design. First, perspective machines applied computation to spatial problems with the view to evolving both artistic and architectural expression. The introduction of linear perspective marked a paradigm shift in art affecting both the techniques and meaning of artistic production. Finally, the computation performed by perspective machines was mechanical; a 'pure' expression of analogue computation (the analogue computers of the 1930s and 1940s would take advantage of electricity and electromagnetic properties of matter whose computation was invisible to human eyes). In other words, perspective machines sit in stark contrast to digital computers and, therefore, allow us to foreground how machines impacted epistemological processes of spatial representation.

Among the plethora of designs for perspective machines, Dürer's incision *Man with Lute* (1523) perhaps depicts the most accomplished example of such machines as the process of drawing a perspectival view of the lute is almost entirely automated (with the exception of the human figure moving a needle along the surface of the musical instrument).Observed from the point of view of the relation between computational techniques and design, perspective machines could be understood as 'built thoughts'. Such machines were shaped after the scientific theories of the time and their implementation in physical artefacts provided artists and architects with new means to apprehend reality. Such experiments had mostly an intellectual rather than practical value with little or no application (in the early Renaissance, the aim

[3] The diffusion of perspective as technique was also accompanied by the publication of more practical manuals that did become part of the skill set that artists made used of. However, such publications emerged at a later stage, particularly in the fifteenth century in Central and Northern Europe.

was not to make the creation of work of art faster). Eventually, plans for more pragmatic applications emerged: for instance, perspective machines had a military use as they allowed the drawing of plans of fortresses without closely surveying them.[4] A general overview of the key features of perspective machines will help us grasp how their computation worked, how they apprehended elements of reality and impacted the design process.

Perspective machines were by all accounts computational devices. Not only did they make calculations, but they also presented some of the basic characteristics of computation such as information compression (selectively capturing certain aspects of phenomena), encoding (through a system of pullies, strings, etc. sight became measurable and computable), and circulation of information (the output of the analogue computation could be translated into a stable medium such as numbers to be reproduced multiple times and, possibly, at different scales). Contrary to modern digital computers, their computation was analogue as they made calculation by measuring lengths and angles. As is the case for mechanical analogue computers, in semiotic terms, perspective machines were icons; more precisely, they were diagrammatic icons as they captured the relations and the idea between the object and its representing sign (Pierce 1998). The way in which perspective machines signified was not based on an arbitrary, conventional pairing between phenomena and signs, rather each material was chosen and parts modelled to best translate in physical form the theories that informed the functioning of the machine. The overall design followed as closely as possible the theoretical model that inspired them; in other words, an informed observer could potentially infer the theory embodied in the device by simply looking at it. Perspective machines (which we now consider as very simple analogue computers) were not cryptic to understand, they had certain pedagogical qualities deriving from the lack of abstraction of their computation, whereas, on the other hand, abstraction forms one of the central tenets of code (symbolic, discrete computation). One of the issues of analogue computation is that iconic signs are in principle compromised by material debasement (Rotman 1994).Perspective machines were no exception as they would not operate correctly, or at all, if materials worn out or connections between parts stopped functioning. Finally, perspective machines could only perform a limited range of computations. Dürer's device could only compute the theory of linear perspective; the model of computation embedded in the machine was not transferrable to other tasks. Even though the analogue computers built in the first half of the twentieth century could generalise their computation to a wide range of applications, the iconic nature of analogue computation always implies a movement toward generalisation which starts with conceiving a device for a specific problem (e.g. how to construct a linear perspective) to then apply it to other computational problems to test its transferability. Computation and materiality are so fundamentally linked in analogue computers that the problem of universality cannot be posed at the outset but only introduced afterwards by extending the application of the analogue

[4] For instance, in *Le Due Regole della Prospettiva Pratica* (1583) Vignola's illustration portrays a perspective machine which is in principle possible to use for surveying larger objects (a statue in the etching) or landscapes.

computer to a broader set of problems. As a result, the issue of transferability is always present when working with analogue computers. The limits of the domains analogue machine can compute also marks the limit of their ability to perform as prosthetic devices; that is, to make an unknown domain intelligible and, in turn, to advance our knowledge of it.

A different kind of paradigm applies to machines propelled by code as in the case of modern digital computers. Code differs from analogue machines as it no longer simulates a particular physical feature but thought itself. If, as Sini (2009) suggests, among others, technology is a prosthesis that extends the faculties of the body to produce knowledge, then code no longer augments any motorial activity, but the mind. The realisation that code is a vehicle for the circulation of thought is perhaps made visible for the first time by the automated Jacquard loom in which the 'intellectual' activity of controlling the machine is handled by code (in the form of punched cards), whereas the 'physical' activity is performed by a traditional loom.[5] Conceived as an abstraction, code is not developed on the basis of material or physical properties as in the case of analogue machines; its roots are in language, particularly in language's ability to represent phenomena through abstraction.

As for analogue computation, code too presents the fundamental qualities of compression, encoding, and circulation. Semiotically, however, code does not belong to the class of iconic signs but rather to that of symbolic ones. Code bears no direct visual or material link to the objects it represents; the issue of material debasement that characterised analogue devices is here absent. The central position occupied by abstraction and the manipulation of symbolic signs meant that the intellectual project to develop code (what Rossi (1960) describes as *Clavis Universalis*) is first and foremost striving for universality. Since Ramon Llull's proto-computational experiments in the late Middle Ages, the ambition has been to conceive a system of signs that could represent anything (Eco 1995). Transferability is not an acquired feature but an essential characteristic of any attempt to generate such system of symbolic signs. The long search for a universal code found a closure in the work of Gödel (1931) and, more importantly for this discussion, Turing (1937). In the Universal Turing Machine (UTM) we find a resolution to a number of concerns that had been animating the search for a universal language and often hindered its progress: the UTM is economic (it employs the smallest set of symbols possible (0 and 1) to represent phenomena), is potentially transferrable to a physical device (the digital computer that appeared shortly after the end of WWII were based on Turing's paper), and, most importantly, is universal (not in the sense that Hilbert had hoped, but rather, the UTM can

[5] The example of the Jacquard loom also allows us to make a further clarification. The characterisation of code as a disembodied concept is not entirely precise as it only accounts for how code is conceived. At its conception code is a pure abstraction which does not strictly need to exist in reality; however, computers are physical devices which require code to be inscribed onto a material support. The gates of a digital computer are the physical equivalent of the 'disembodied' 0s and 1s of Boolean logic and so are the perforated cards controlling the weaving patterns of a Jacquard loom. As Aden Evens suggests, it follows that at their very core digital computers are still analogue machines: 0s and 1s are arbitrarily assigned to, for instance, fluctuations in voltage whose variation is continuous, not discrete (Evens 2015).

compute anything that can be computed). Code and, by extension, the contemporary digital computer bear no pedagogical qualities. The arbitrariness of symbolic signs removes the possibility of extracting meaning by observing its functioning. The labour of extraction and circulation of knowledge in digital computation is therefore a central issue: the task is to translate the abstract, cryptic language of symbolic computation into an intelligible one, (and vice versa). The aim is to test the capacity of symbolic language to represent and simulate empirical phenomena.

The comparison articulated above shows the historical roots of cryptography by demonstrating how central an issue it has been for computation as well as how issues of cryptography vary between analogue and code-based machines. Far from being concerns belonging to a distant past, the current implementation of ML methods in design reproposes similar issues involving the construction of a communication channel between human and machine intelligence. How does knowledge emerge from an incomprehensible—for humans—operation performed by learning algorithms? What role and agency will automated operations performed by ML algorithms have on design processes? The problem of knowledge in architecture therefore is reformulated as the design of systems of the exchange between different domains: that of 'meaningless' strings of symbols of code, their intelligible translation which becomes actionable, and the discipline of architecture. The penetration of Artificial Intelligence (AI) in architectural and urban design makes cryptography a central concern as designers and machines are constantly performing operations of translation between diverse domains, syntax and semantics, multi-dimensional and three-dimensional representation (Fig. 7.2).

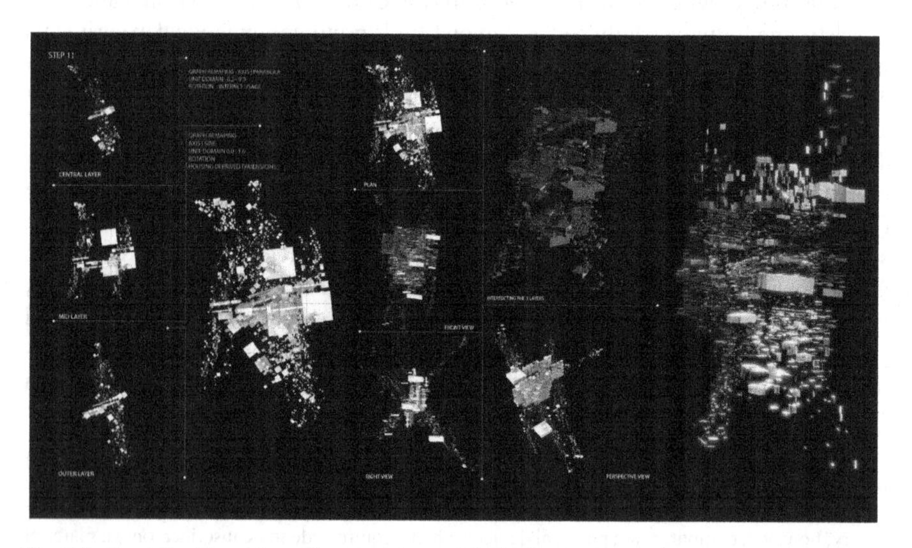

Fig. 7.2 Operative Rewritings. Exercises in spatial distribution based on large datasets. Author: Flora Mistica Selvaraj, B-Pro Urban Design, Research Cluster 14 (R. Bottazzi, T. Varoudis, E. Tsouknida), 2020–21

7.4 Cryptographic Computation

As we have seen, one of the central issues of symbolic computation is that of translation. Cryptography is the specific branch of computational studies tasked to devise methods for the exchange between different types of signs: more precisely, between random signs (cypher text) and intelligible ones (plain text). Beyond its historical association with diplomacy, cryptography can be understood as a process of extracting knowledge from previously incomprehensible domains. How does this process work in the case of ML methods applied to architectural design?

Though it is well beyond the scope of this paper to revisit the history of cryptography, one particular instance, the polyalphabetic cypher conceived by Leon Battista Alberti, provides a significant historical example to grasp what cryptographic methods offer in the exploration of complex, 'meaningless' sets of signs. The idea of a non-mimetic representation which departed from depicting the sensible appearance of objects had already been part of Western culture and already found a proto-computational manifestation in the work of Ramon Llull. Llull's system computed letters by spinning a set of concentric wheels which returned strings that could be interpreted by using Llull's own tables. The precedent is relevant because cryptographic methods were deployed to unlock the hidden secrets of cosmos rather than safeguarding the secrecy of diplomatic cables. Mathematics supplied a grammar for the system in the form of combinatorial logic. As Leibniz also noted, the potential of combinatorial thinking was however hindered by Llull's pre-conceived metaphysics which forced him to eliminate many of the combinations possible. In other words, combinatorial logic was not understood as a genuine instrument for search, but rather as a method to confirm postulated truths. Despite the limited use of mathematics, Llull's combinatorial logic already indicated the possibility to employ mathematical thinking 'as a way to *work* and to actually investigate and 'compute' the world' (DuPont 2018, p. 101). In other words, combinatorics performed the labour of extracting and mobilising knowledge 'stored' in nature, as an instrument to inquire into nature's deeper mechanisms and true meanings.

Part of the conceptual limitations of Llull's system was overcome in Lean Battista Alberti's work on cryptography whose notational system employed mathematics as a searching instrument. Described by historians as the first modern cryptographic method ('a new species' in Kahn's (1967) words), *De componendi cifris* (1466 ca.) illustrated Alberti's (1998) polyalphabetic cryptographic system, the first of its kind, which significantly improved on the security of the encryption by multiplying the number of cyphers and varying them throughout the encoding process. The device was formed by two concentric rings (very similar to Llull's) each divided into cells containing the letters of the alphabet (Kahn 1980): for each letter in the outer ring, the user would look up the corresponding letter on the inner one. Mathematically, Alberti's polyalphabetic method made decryption by frequency analysis much more difficult and therefore showed greater 'statistical awareness' about the application of mathematics to language by acknowledging that letters did not distribute randomly but rather are statistically dependent events. Also, polyalphabetic encryption exponentially grew the size of the space of all possible combinations making such space only surveyable through mathematical means; mathematics became a prosthetic device to articulate reasoning. Alberti's method applied mathematical instruments on language by constructing a notational system which could be manipulated in abstract terms and potentially transferred to other domains to determine what data to select, store, and process. In other words, Alberti's notational method established a communication channel between strings of 'meaningless' signs and intelligible ones. All the mathematical operations performed on the cypher text to decrypt it were pre-semantic: they applied syntactical transformations (based on character substitution) without complying with a pre-established meaning. The system operated syntactically as it was able to 'select an array of marks from the noisy reservoir of all possible constellations' (Wellbery 1990, p. xii). Historians of sciences point out that, since their inception in the Renaissance, notational systems expanded well beyond cryptography to be deployed to decrypt the 'Book of Nature' and, as Francis Bacon noted, exploit the power of cyphers to reverse-engineer nature itself with the aim of connecting 'anything by anything' (Bacon 1962, p. 139).[6] A notational system in fact cannot be mimetic as one of its conditions of existence is precisely to guarantee a difference between the system of signs proposed and the phenomena encoded. The severing of the intuitive, visual link between cypher and plain texts brings into play notions of transposition, substitution, and re-writing which, in turn, can connect it to other abstract systems of signs, thus opening up a speculative use of cyphers and cryptographic methods in general. Such quality would be consistently exploited by the philosopher of the renaissance to eventually form the basis of algebraic mathematics at the beginning of the seventeenth century and constitute the basis of the work on computation of Leibniz, amongst others. The central figure in these operations of translation is the cypher as the device that tunes two domains to each other: one domain constituted by random, 'noisy' signals, whereas the other is characterised by patterns and the emergence of intelligible figures (Fig. 7.3).

7.5 Cyphers

Cyphers are understood as rule-based mechanisms to hide (encrypt) or reveal (decrypt) messages.[7] Semiotically, however, cyphers are special types of signs. Rotman (1994) defines them as meta-signs because they both belong to and organise the set of signs they are part of. In Alberti's system the cypher was a letter (or a sequence of them given the polyalphabetic nature of his cryptographic approach) that encoded other letters: the cypher was a special member of a set which both gave rise to and foregrounded the relationships between all the members of the set.

[6] "Omnia per Omnia" in the original Latin.

[7] "An algorithm for encryption or, in its inverse form, for decryption" (Daintith and Wright 2008).

Fig. 7.3 Operative Rewritings. Exercises in spatial distribution based on large datasets. Author: Alankrita Amarnath, B-Pro Urban Design, Research Cluster 14 (R. Bottazzi, T. Varoudis, E. Tsouknida), 2020–21

Because of their semiotic properties, cyphers fulfil both conditions outlined in describing how prosthetic objects work. First, they can be added to a set and, in turn, they will expose connections between the members of such set. Cyphers, in fact, are particular signs whose eminent value is that of structuring other signs by making them intelligible. The process of translation is also one of production: the application of the cypher generates a new set of signs amenable to further manipulations. In ML, the application of a dimensionality reduction algorithm[8] to a complex dataset broadly follows an analogous process: the algorithm surveys the dataset to return statistical correlations between its members. What emerges here is not only the generative capacity for cyphers to instantiate the production of signs, but also

[8] The reduction of the number of dimensions in a dataset is one of the central challenges posed by working with Big Data. Digital designers working with large datasets are bound to encounter this issue at some point in their work. In data science dimensions do not have a spatial meaning, rather, they identify the number of features describing each data point. For instance, though never fully confirmed, it is rumoured that each Facebook user is described by 52,000 types of data (Green 2018). A hypothetical spreadsheet of all Facebook users would therefore consist of 1.69 billion rows (approximate number of Facebook users at the time of writing) and 52,000 columns or dimensions. Due to the impossibility for human cognition to navigate such a vast data space, computer scientists developed dimensionality reduction algorithms whose function is to diminish the size of the dataset (both in terms of number of features and, potentially, data points) whilst minimising the loss of information contained in the original dataset. There are many different algorithmic procedures available to reduce the number of dimensions of a dataset, however all the literature on the subject concedes that the reduced dataset will always only be a partial representation of the original (though the reduced dataset will capture the essential qualities of the original one). Dimensionality reduction is one of the most common and debated procedures in data science; as such, it is an issue that anybody working with Bid Data eventually confronts in their work. For these reasons, dimensionality reduction algorithms provide a good testing ground to analyse and prove the notion of the cypher. See both 'Dimensionality reduction'. *Wikipedia.* Available at: https://en.wikipedia.org/wiki/Dimensionality_reduction (accessed 6 December 2021) and 'Feature (machine learning)'. *Wikipedia.* Available at: https://en.wikipedia.org/wiki/Feature_(machine_learning) (accessed 1 January 2022).

the speculative character of this operation. Projected onto the noisy, alien domain of signs, cyphers guide their exploration by following clear rules and return intelligible, relational patterns rather than fixed meaning or truth. Cyphers can remap speculative thoughts onto data, test their viability as structuring thoughts in order to elicit further analysis. From the point of view of prosthetics, the epistemological work of the cypher is always incomplete and additional inputs are required (other datasets, counter-factual hypothesis, testing, etc.). The agency of the meta-sign of cyphers enables us to perform thought-experiments about the systems of signs we are confronted with, mobilise them (or put them to work, in Quinn DuPont's characterisation of Alberti's cryptographic system[9]). The labour of designers consists in keeping the process of exploration developing, providing different frames of reference to contextualise and instrumentalise the strings of signs produced by cyphers; the disciplinary discourse provides a rich catalogue of topics to integrate and critique the algorithmic explorations such as issues of scale, programme, distribution, etc.

7.6 Cyphers and Spatial Design

The long journey through the history of computational techniques and their impact on design helped us better grasp the central role played by techniques in the architectural discourse and how ML methods can impact them. Since the 1960s, architects have favourably looked at the development of design models based on systematic thinking which emerged both in the fields of cybernetics and linguistics. Such a move was triggered as a response to the crisis of Modernism which, by then, was perceived as a sterile movement that had lost its innovative charge. The search for alternative models concentrated on procedural techniques that shifted the attention away from formal solutions in favour of approaches that granted design processes a prominent status. Parallel to the research carried out by artists and semioticians, architects established indexical relations between the design process and formal outcome in which traces of the former could be legible in the latter. Allen (2006) charts the role that Peter Eisenman had in foregrounding process in design, but also extends to the fact that the first generation of digital architects in the 1990s still operated within the path the American architect had traced.Though digital architects in the 1990s could take advantage of new techniques provided by advanced software (morphing, animation, etc.), the formal outcomes were still legitimised by their ability to bear the marks of the process that generated it. At the core of the debate was once again the persistent impossibility for architecture to regulate or even fix the relation between signs, meaning, and interpretation; an issue that is still current and that a cryptographic approach to digital design can provide new insights on.

[9] DuPont (2018). *The Printing Press and Cryptography: Alberti and the Dawn of a Notational Epoch.* In: Ellison and Kim (2018), *A Material History of Medieval and Early Modern Ciphers: Cryptography and the history of Literacy*. Routledge.

The figure of the cypher can be of relevance in discussing digital techniques as it offers a different model for articulating process and product (or, better, data and form) by moving away from both the iconic computation of perspective machines and the indexical one of procedural design. Contrary to indexical signs, cyphers in fact do not bear direct marks of the objects they are representing. In fact, it may not even be appropriate to speak of representation as cyphers do not work mimetically—be it in an iconic or indexical fashion. Rather, cyphers survey a given domain of data as meta-signs; that is, they restructure data to give it intelligibility, albeit partial. The relation between cyphers and data is mediated by mathematics, the language expressing both the behaviour of the cypher and guaranteeing the incompleteness of the epistemological process initiated. Cyphers generate a new image of the domain surveyed, they open up new opportunities to speculate about space and its articulation. It is this novel image produced by cyphers that allows us to venture beyond the reading of cyphers as prosthesis. Here, cyphers are speculative instruments to articulate new possibilities for design. Magnani (2019) helps us grasp such new possibilities by emphasising the abductive capacity of computational thinking: "computational programs that execute various kinds of hypothetical reasoning can be seen as prosthetic 'abducers': just as microscopes are technologically created to extend human cognitive capacities and methods". Beyond strict formalisation and classical logic, cyphers open up new domains of research for computation to operate within dynamic, open, uncertain, fallible forms of rationality. For instance, hypothetical assertions such as those presupposed by abductive methods or fallible, paradoxical thinking such as those proposed by New Logic (Magnani 2013).Again, we are confronted with an extension of cognitive abilities that allow us to chart, learn, and eventually interact with unknown domains such as those of massive, multi-dimensional datascapes.

Computationally we can draw an analogy between cyphers and algorithms based on their symbolic computation, their clear syntax based on mathematics, and the structuring, relational, and non-indexical quality derived from being meta-signs. As such, cyphers are designed objects as they must be invented and require a great deal of artistry and playfulness whilst solidly resting on strict rules that define them. Cyphers capture elements of randomness of an unknowable dimension through rational operations. In designing with ML methods, the issue of instability does not have so much to do with the relation between signs and their meaning; rather, more radically, with the intelligibility of numerical and statistical representation generated by the application of learning algorithms to massive datasets. No longer bound by iconicity or indexicality, algorithms become speculative instruments to structure new relations. 'Instrumentality is not a means to an end, but a method or a knowing-how tending towards the determination of this or that result' (Parisi 2019, p. 43). The challenge in applying ML algorithms to urban design therefore consists in: on the one hand, a representation of urban complexity through data, and, on the other, the re-articulation of data through automated algorithmic processes which return intelligible and partial representations amenable to further manipulation. Whereas the digital discourse in architecture has grown accustomed to thinking of processes either in a top-down or bottom-up fashion, structuring spaces through data makes such categorisations irrelevant as both conditions simultaneously exist. For instance, classifying algorithms survey massive datasets with unprecedented granularity (e.g. kernels in Neural Networks can extract features in squares of few pixels), without the need for introducing any pre-determined hierarchy of scale. Likewise, the re-structuring of geo-located data performed by clustering algorithms returns novel arrangements of scale, contiguity, similarity, and distribution which can no longer be accounted for through parametric thinking. We can see how the introduction of ML methods in design no longer concentrates on the issue of form as a way to reconcile differences, but rather on notions of relation, structure, transposition, interactivity, openness and incompleteness. In other words, the challenge presented by ML algorithms goes deeper than simply introducing new techniques to affect more profound, historical preoccupations of the discipline.

It is worth revisiting the initial considerations on the semiotics of analogue and coded computation outlined earlier in the text to better grasp what cypher can offer to conceptualise the use of ML methods in design. In discussing the semiotics of photography, Roland Barthes proposed a relation between analogue and coded representation that is relevant for this discussion. Barthes spoke of photographs (an analogue medium) as 'messages without a code' as, contrary to other media such as drawing, they promised a direct, unmediated image of reality.[10,11] The decoding of a drawing rested on the simultaneous presence of three elements which needed not to be there in the case of photography. The translation of drawings implied the presence of: a structure of rules that guided the process of transformation, a gap between reality and its transformation (a drawing is a selective representation of the original scene or object), and an operator that possessed an adequate level of literacy to successful 'encode' the message. In analysing the functioning of perspective machines, we argued that the iconic type of computation of analogue machines had direct, almost pedagogical qualities. Barthes reinforces such reading by extending it to all analogue modes of representation; in the case of photography, the unprecedented level of realism afforded by the medium delivered an unmediated message, so much so that Barthes affirmed that photography did not represent reality by transforming it (as in the case of drawing), but simply by registering it (1985, p. 33).

Design with ML algorithms implies an inversion of the paradigm introduced by photography that allows us to speak of codes without a message. To stress the importance of code allows us to foreground the arbitrary nature of algorithmic operations.

[10] Barthes' ideas are taken from two essays in which the notion of the message without a code was first introduced (Barthes 1977, 1985).

[11] This passage is quoted in the second of two essays Rosalind Krauss dedicated to the artistic production of the 1970's in which she took Barthes' uncoded message and aligned it with the indexical operations of minimal artists. The impact of these essays was not limited to art circles and affected architectures ones as well. Stan Allen convincingly describes the impact that this essay and the general interest in procedural works had on Peter Eisenman's work. It is also important to point out that Eisenman's production at the time should not be understood as a translation of Krauss' ideas, but rather as a body of work already possessing its own trajectory and being influenced by discussion in parallel fields (Allen 2006).

Data, similar to photography, is often understood to emerge from a process of registration rather than representation. However, data can only speak if encoded, that is, formatted in order to be parsed by algorithms (cyphers) regulated by mathematical and semiotic qualities. Cyphers belong to notational (code-based) modes of representation, however, their meta-semiotic qualities allow them to operate on data in ways that greatly differ from previous media. Meta-signs give rise to and make sets of signs intelligible; that is, the design of the cypher determines both the rules of emergence and 'projectability' of the cypher: how it will survey a dataset, how it will structure it and the relations that it will establish between data. Moreover, such rules of projectability are not static as they change throughout the process of decryption (similar to Alberti's polyalphabetic cryptography). Though a great number of complex decisions are involved in the design of ML methods, we can speak of codes without a message because cyphers still operate at a pre-semantic level concentrating on relations (or correlations) that are not yet locked into specific meanings yet. Cyphers provide a special vantage point which returns a precise and yet incomplete representation of a previously meaningless, 'noisy' sets of signs. The operations performed by the cyphers make data intelligible, that is, amenable to further interaction with other variables, examples, and criteria without pre-determining their meaning. In the case of urban data, ML methods foreground aspects or relations of and within urban environments which were cognitively inaccessible before. For instance, in the case of dimensionality reduction algorithms, such class of algorithms allows designers to engage with cognitive impenetrable datasets, in order to return insights on the domain surveyed which solely result from algorithmic operations. The comparison between the same data set distributed according to its original geolocation or the abstract space of dimensionality reduction algorithms immediately shows that typical hierarchies structuring space (such as top-down or bottom-up) no longer apply as spatial distribution is rather characterised by discontinuities, jumps, heterogeneous zones. Received notions of site, scale, organisation no longer apply and new methods of interventions are needed. For these reasons, working with ML methods can be described as working with codes, that is cyphers which dynamically survey a complex, noisy space which escapes semantic determination (Fig. 7.4).

7.7 Conclusions

ML methods will not only quickly diffuse in architecture and urban design, but also, and more importantly, they will challenge established models for representation of space in ways that will demand greater theoretical and intellectual attention from designers. The prospect of speeding up design by making it more efficient will be a short-lived gain which may focus on the most immediate and superficial benefits of ML whilst obscuring more profound transformations related to space and its manipulation. What is at stake in the introduction of ML methods in architecture is to interact with forms of automated logic that are truly different from human cognition and demand architects to design systems of communication and exchange between

Fig. 7.4 N^2P^2—Neural Network Public Places. Project for the reorganisation of the Île de la Cité in Paris. Perspectival section through the Île de la Cité the Authors: R. Bottazzi, T. Varoudis, P. Prajapati, X. Wang

human and algorithmic thinking. The reward in taking on such a challenge is to open design up to a new logic able to operate at multiple scales, be tested against different conditions, and, through data, be able to provide agency to factors that have been historically marginal, or even absent, in the design process. From climate change to new types of spatial organisation and collaboration, the list of practical applications of ML methods in design provides a rich and timely set of issues for architects to experiment with.

References

Alberti LB (1998) Buonafalce A (ed) De componendis cyfris (1466). Galimberti Tipografi Editori, Turin

Allen S (2006) Trace elements. In: Davidson C (ed) Tracing Eisenman—Peter Eisenman complete works. Thames&Hudson, London, pp 49–65

Bacon F (1962) The advancement of learning. Edited with an introduction by G. W. Kitchin, 1st edn published in 1605. Dent, London

Barthes R (1977) The photographic message. Image music text (trans: Heath S). Fontana Press, London, pp 15–31

Barthes R (1985) The rhetoric of the image. In: The responsibility of forms: critical essays on music, art and representation (trans: Howard R). Hill and Wang, New York, pp 21–40

Carpo M (2017) The second digital turn: design beyond intelligence. The MIT Press, Cambridge, MA
Clark A, Chalmers D (1998) The extended mind. Analysis 58(1):7–19
Daintith J, Wright E (eds) (2008) Combinatorial explosion. In: A dictionary of computing, 6th edn. [online]. Oxford University Press, Oxford. https://www-oxfordreferencecom.libproxy.ucl.ac.uk/view/10.1093/acref/9780199234004.001.0001/acref-9780199234004-e-857?rskey=LRpMhl&result=959. Accessed 5 Feb 2023
DuPont Q (2018) The printing press and cryptography: Alberti and the dawn of the notational epoch. In: Ellison E, Kim S (eds) A material history of medieval and early modern cyphers. Routledge, New York, NY, pp 95–117
Eco U (1995) The search for the perfect language. Blackwell, Oxford
Evens A (2015) Logic of the digital. Bloomsbury Academic, London
Gödel K (1931) Über formal unentscheidbare Sätze der Principia Mathematica und verwandter Systeme, I. Monatshefte Math Phys 38(1):173–198
Green A (2018) Facebook's 52,000 data points on each person reveal something shocking about its future. https://www.komando.com/social-media/facebooks-52000-data-points-on-each-person-revealsomething-shocking-about-its-future/489188/. Accessed 5 Feb 2023
Hayles NK (2014) Cognition everywhere: the rise of cognitive nonconscious and the costs of consciousness. New Lit Hist 45(2):199–220
Kahn D (1967) The codebreakers: the story of secret writing. Macmillan, New York, NY
Kahn D (1980) On the origin of polyalphabetic substitution. Isis 71(1):122–127
Kemp M (1990) The science of art: optical themes in western art from Brunelleschi to Seurat. Yale University Press, New Haven & London
Magnani L (ed) (2013) Introduzione alla New Logic: Logica, filosofia, cognizione. Il Melangolo, Genoa
Magnani L (2019) AlphaGo, locked strategies, and eco-cognitive openess. Philosophies 4(1):8
Parisi L (2019) The alien subject of AI. Subjectivity 12(1):27–48
Pierce CS (1998) What is a sign?. In: The essential pierce. Selected philosophical writings, vol 2 (1893–1913). Indiana University Press, Bloomington, IN, pp 4–10
Rossi P (1960) Clavis Universalis: Arti Mnemoniche e Logica Combinatoria da Lullo a Leibniz. R. Ricciardi editore, Milan
Rotman B (1994) Signifying nothing; the semiotics of zero. Stanford University Press, Stanford, CA
Sini C (2009) L'Uomo, La Macchina, L'Automa: Lavoro e conoscenza tra future prossimo e passato remoto. Bollati Boringhieri, Turin
Turing AM (1936–1937) On computable numbers, with an application to the Entscheidungsproblem. Proc Lond Math Soc s2–42:230–265; correction ibid. s2–43:544–546 (1937). https://doi.org/10.1112/plms/s2-42.1.230 and https://doi.org/10.1112/plms/s2-43.6.544
Wellbery DA (1990) Foreword. In: Kittler FA Discourse networks 1800/1900. Stanford University Press, Stanford, CA

Chapter 8
On Legibility: Machine Readable Architecture

Andrew Witt

8.1 Introduction

As contemporary formats of spatial data proliferate—image sets, 3d scans, and geospatial databases, among others—the question of *how to read* these various media architecturally becomes a critical theoretical and practical matter. The question is not only a natural response to the vast new aggregations of digital architectural information with which we are confronted, but also a query into the fundamental relationship between human and machinic forms of perception, classification, and reading. Architects and machines are reading the digitized representations of architecture in increasingly imbricated and symbiotic ways. At the same time, new ways to derive meaning from spatial datasets, such as the training artificial intelligence, machine learning, and machine vision processes, allow the tastes and judgements of human visual intuition to be encoded and externalized in software tools with uncanny nuance. Such techniques interlace human vision and digital scanning in strange new ways and prompt reflection on the very act of reading architecture (Fig. 8.1).

This essay unpacks the oscillating valences and expanding contours of "readability" as they pertain to machinic encodings of data on the one hand and the visual language of architecture on the other. Three distinct narratives offer parallel interpretations of the entangled practices of architectural and machine readability. The first is a capsule chronology of machine readability, from its roots in tabular statistical datasets in the nineteenth century to its convergence with AI and machine learning today, with a moment of critical inflection in the 1960s and 1970s. The second is a complimentary examination of the humanistic concept of architectural readability, particularly the legibility of graphic drawings, as it evolved since the 1970s. The third explores the intersections of machine and architectural readability through two

A. Witt (✉)
Harvard Graduate School of Design, Cambridge, MA, USA
e-mail: awitt@gsd.harvard.edu

P. Morel and H. Bier (eds.), *Disruptive Technologies: The Convergence of New Paradigms in Architecture*, Springer Series in Adaptive Environments,
https://doi.org/10.1007/978-3-031-14160-7_8

Fig. 8.1 MTS_003, an architectural sculpture generated by the classification of irregular elements using machine vision. Certain Measures, 2021

projects that use machine vision as a means to extend the creative and interpretive power of architects.

Across these three narratives, reading emerges as a peculiar process of piecing together of fragmentary marks and ciphers into coherent wholes, a process akin to the creative act of architecture itself. Beyond exploring the idea, lineage, and cultural antecedents of machine readability, this paper shows what machinic legibility means today through new practices of reading form and generating architecture. From projects that morphologically catalog the world's billion buildings to the application of shape classification for radical waste reuse, machinic reading is transforming the roles and products of design. As machine intelligence is sensitized to recognize the signs, qualities, and language of architecture, it can aid architects to map and recombine them not only in new forms but also new epistemic regimes.

8.2 Machine Readability

Reading, as a human and machine practice, aligns perceptual, cognitive, and linguistic faculties in a series of discrete operations that translate visual marks into meaningful interpretations. In the context of machine readability, one sees the perceptual task of scanning—the automatic transcription of visual elements into an encoded and processable form—as an essential starting point. Reading and scanning evoke different epistemic valences. Reading is a relational and synthetic activity that associates sequences of symbols in particular interpretive configurations with complex meanings. Scanning, in contrast, is an act of transcoding data from one form to another, from optical to alphanumeric, or from marks into indexed letters, for example, which the historian Zeynep Çelik Alexander has admirably recounted.[1] Reading often presupposes scanning, particularly in a machinic context.

The history of machine readability follows an inexorable expansion of the types of data which can be read automatically. From the highly encoded and explicitly structured tabular data necessary from the late nineteenth century to the loosely encoded and unstructured imagery or scans that are currently possible, the scope of machine readability has gradually increased. When the curious phrase "machine readable" appeared around the 1920s in the United States, it referred narrowly to numeric records that could be easily and efficiently tabulated by electromechanical means. Census summaries, accounting records, and inventory ledgers were examples of the serialized datasets that could be machine readable. Written documents in natural language—literary texts or even more mundane alphanumeric lists—were well beyond the scope of machine readability, let alone the visual shapes and forms of architecture. Machine readability presupposed data that had been transcribed into a particular format—such as punchcards or ticker tape—for automated electromechanical consumption and processing. While humans might read text, images, situations, or even other people, machines read numbers or, perhaps more fundamentally, abstract sequences of sparse dots. Machine readable data was numeric and discretely encoded.

By the 1960s, machine readability reached an inflection point as it expanded to encompass not only numeric encodings but also various encodings of text and even the organization of that text. In fact, it began to entail the organization of knowledge through the creation of vast card catalogs or indices of text. Machines not only tabulated data but also digested and collated it, rendering it navigable and interrelated. A key appearance of the term "machine-readable" in this period was in IBM researcher Hans Peter Luhn's 1958 article "The Automatic Creation of Literature Abstracts." Luhn attempts to substitute the "intellectual effort" as well as "human effort and bias" required to read and summarize a complex text with automated sequences of statistical metrics such word frequencies.[2] Key to Luhn's approach was the automatic

[1] Alexander (2020).

[2] Luhn (1958a).

identification of "significant words" that betray the meaning of larger phrases and texts.[3]

> [T]he method to be developed here is a probabilistic one based on the physical properties of written texts. No consideration is to be given to the meaning of words or the arguments expressed by word combinations. Instead, it is here argued that, whatever the topic, the closer certain words are associated, the more specifically an aspect of the subject is being treated.[4]

A statistical method of extracting topical or thematic information echoes remarkably current computational methods of natural language processing. In Luhn's description, we glimpse the shape of things to come in machine readability: a search for formal clues and semantic markers which would provide insight to underlying meaning (Fig. 8.2).

New categories of pattern and feature recognition broadened the purview of machine readability to include more complex graphic images. By 1963, inventors had begun to develop idiosyncratic contraptions that transcoded human-made marks such as handwriting into machine-readable representations. For example, Dimond's 1963 patent for "Machine Reading of Handwritten Characters" deployed a "translator" to scan and rectify the optical impression of written text.[5] Around the same moment, medical researchers began to speculate on the "graphical and pictorial information" in x-rays for evaluation and diagnosis.[6] Machine readability thus expanded beyond a primarily symbolic process to an interpretive one—the divination of inflected nuance from images, diagrams, and other graphic content. This expansion to graphic content foregrounded "feature recognition," a way station between scanning and reading. Feature recognition first emerged as a desirable machine facility in the 1960s, a complement to pattern recognition. First applied to decipher physiological imagery, it was useful more broadly as a class of automatic sensemaking.[7]

Though the power and speed of these methods of automatic scanning and machinic reading advanced, they were always supported by explicit and programmatic methods of evaluation. Unusual or nonconforming input often confounded such processes. Moreover, the processes themselves had to be understood in extreme detail in order to be explicitly programmed. Machine readability presupposed a rigid transcription of unstructured content into coded and structured data before it could be read. Even the graphical and pictorial information consumed for medical imaging was parsed in explicit and regimented ways, with little margin for deviation or flexibility.

The past two decades have brought a second inflection point in the flexibility and generality of machine reading processes. Reliance on explicit methods of feature recognition has given way to more flexible techniques that uncover implicit patterns in unstructured data through large-scale statistical correlations and trends. New varieties of neural nets and machine learning have upended that regimented notion of reading

[3] Luhn (1958b).

[4] Luhn (1958b).

[5] Dimond (1963).

[6] Taylor (1960).

[7] Talbot and Harrison (1966).

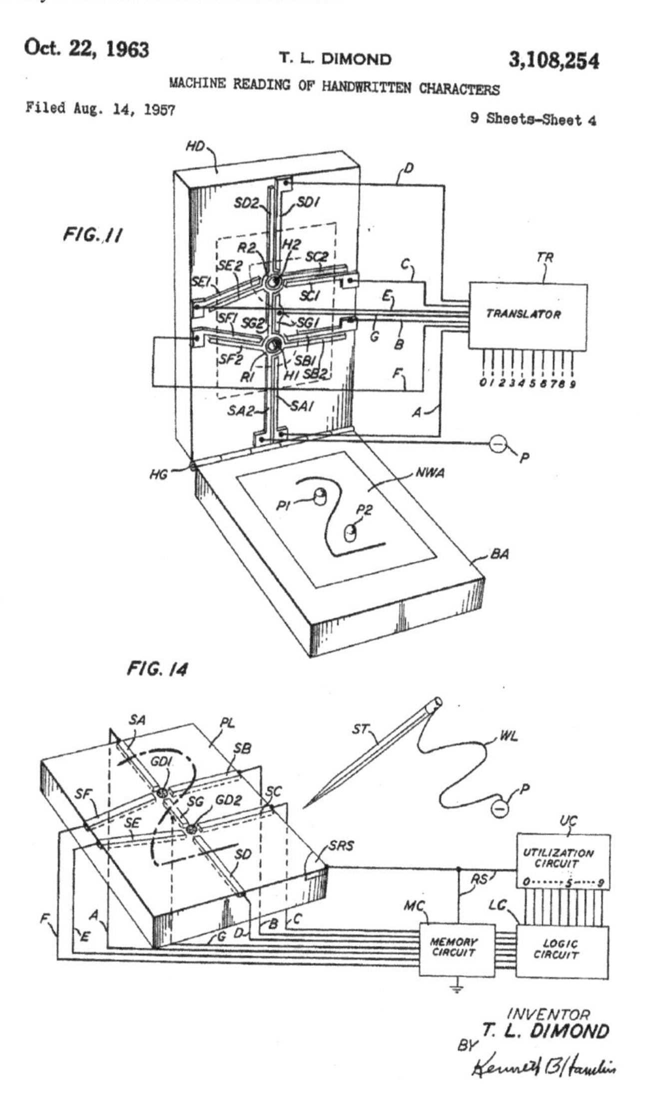

Fig. 8.2 A 1963 device for "machine reading," a method for quasi-optical character recognition. U.S. Patent Office

data and, by extension, architecture. Trained algorithms can now read and organize vast troves of unstructured data automatically, from architectural photographs to building plans, from construction waste to the verbal descriptions of shape. Neural methods extract tacit, even hidden relationships from the arrangement of various visual cues. By dramatically amplifying the visual intuitions of designers, machinic classificational methods have become more akin to search engines, transforming once unwieldy data into creative insight.

By scaling up the process of formal reading, neural and machinic methods can assemble new searchable catalogs or repositories of material, shape, and form. Of course, catalogs of the elements of design have long shaped the methods and intuitions of architects, from Sebastiano Serlio's catalogs of classical form to Jean-Baptiste Rondelet's enlightenment manuals and beyond. Catalogs offer training and literacy in formal archetypes that exemplify the discipline of architecture. Implicitly, this classification rests on proportions, geometry, topology that are now machine readable.

At its core, reading is the interpretation of marks. The figural polygonal forms of buildings—the silhouettes of their floorplans and elevations, or the shapes of individual architectural elements—comprise a specific set of marks analogous to alphabetic ciphers and equally amenable to machinic reading. Thus, from a technological point of view, today's machinic reading of architecture would seem poised to open new avenues of design experimentation.

8.3 Architectural Readability

Any theory of machine readable architecture must come to terms with the long humanistic tradition of reading architecture. The notion of the building as a document legible in quasi-linguistic ways has enjoyed currency since at least the nineteenth century. Architectural critic John Ruskin, for instance, argued that criticism of buildings and texts were analogous, and traces the idea itself to Quatremère de Quincy's argument that Egyptian monuments, encrusted with inscriptions, were literal texts.[8] But reading architecture as a humanistic theoretical practice and as a system of cultural signification truly came into its own in the 1970s, even as machine readability matured apace. Mario Gandelsonas and David Morton's 1972 essay "On Reading Architecture" is a compelling candidate for the ur-text of this movement. In this essay they argued that "the system of architecture is a system of cultural meaning," and architectural configurations were amenable to a quasi-linguistic method of either syntactic or semantic analysis. Ironically, Gandelsonas and Morton's attention to reading was catalyzed by their fear that "linking architecture to computer technology and sophisticated mathematical models…tends to shift architecture further into the realm of engineering."[9] Against this tendency was their view of "architecture as a system of cultural meaning; it attempts to explain the nature of form itself, through viewing the generation of form as a specific manipulation of meaning within a culture."[10] Reading architecture was thus a foil and antidote to computation. Drawing on the then-fashionable structuralist models of French anthropologist Claude Lévi-Strauss and semiotician Ferdinand de Saussure, they contended that architecture was a "system of significance" similar to written or spoken language. Gandelsonas and Morton's paper was one of the first signals of an evolution of the

[8] Forty (2004).

[9] Gandelsonas and Morton (1972a).

[10] Gandelsonas and Morton (1972a).

understanding of architectural legibility from merely the interpretation of graphic markings to a meta-system in which to evaluate the cultural significance of buildings and architectural projects.

For Gandelsonas and Morton, the practice of reading architecture rested on formal analysis of its constituent graphic and tectonic elements. They identify two divergent strains of such analysis: a semantic tendency, concerned with the cultural meaning of signs then exemplified by Michael Graves, and a syntactic tendency, interested in the abstract relation of signs among themselves and exemplified by Peter Eisenman. Their assessment of the semantic strain was particularly intriguing in its dependence on a repertory or repository of architectural ideas or elements—fragments, quotations, references that collectively mark a larger network of meaning:

> Architectural form can be seen as the manifestations of the codes, plus "quotations" drawn from the architectural repository. These quotations are sets of ideas, images in general, and notions about buildings in particular. In drawing from this repository, the architect can select any form or idea in its original state; he can use formal patterns directly from the five orders of Classical architecture, for instance, or he can use aspects of Mediterranean popular architecture as found in Le Corbusier.[11]

To the modern reader, the idea of the architectural repository curiously echoes the contemporary notion of dataset: a collection of discrete elements from which to compose larger assemblages. In fact, it is this repository-like aspect of architectural content that ironically allows the application of data science methods to architecture.

Though it was only one manifestation of a "linguistic turn" which had reshaped many fields by that point, Gandelsonas and Morton's argument was particularly illuminating for the computational bogeyman that it set up in contest with architecture itself. Computational methods were *ipso facto* proof of elicit and suspect engineering associations with architecture. Yet Gandelsonas and Morton may not have considered how machines might read architecture and thus transform their avowedly humanistic notion of architectural reading. Over time, the capacities of machine to read that once seemed insufficient for or incommensurate with architectural readability began to improve and ultimately approximate the human reading of architecture in compelling ways. As architectural and machine readability converge, the critical possibilities of reading as a generative computational process begin to reveal themselves.

8.4 Measuring Architectural Language

The distinction between architecture-as-language and a more calculational disposition towards design long seemed self-evidently antithetical. In fact, architectural theorist Branko Mitrovic argued that the resurgence of formal geometric interest in the 1990s was incompatible with Gandelsonas and Morton's more referential architectural reading and effectively overthrew it.[12] Yet far from seeing the linguistic and

[11] Gandelsonas and Morton (1972b).

[12] Mitrovic (2009).

geometric tendencies as in conflict, the convergence of human and machinic reading recasts them as symbiotic and complementary. In fact, new modes of machine reading suggest ways of transcending the semantic/syntactic dichotomy with more synthetic kinds of automatic sensemaking.

Today, the disparate practices of machine and architectural readability are interweaving in surprising ways. In adjacent disciplines like the digital humanities, Lev Manovich and Franco Moretti propose ways of reading data in aggregate that illuminate cultural issues and trends, and in some cases rely on geometric or image-centric computational analysis.[13] More humanistic approaches to data science, from sentiment analysis to relational metrics of meaning in natural language processing, are revealing that many qualitative nuances of language and culture can be quantitatively metrized. Subtleties of language can even be geometrized through neural techniques like Word2Vec, so that subtle semantic proximity is mapped into spatial distance. In a similar way, neural and machine learning methods of image classification have opened possibilities for embedding judgements of taste in wider quantitative frameworks.

Machine readable architecture builds on these advances in the quantification of qualities to make the nuance of architecture computationally legible. To illustrate this potential, we present two projects that transform unstructured visual data into meaningful architectural categories—building elements and types. As a point of departure, these projects begin by scanning and classifying simple and elemental two-dimensional shapes or marks akin to silhouettes. Silhouettes occupy a distinct place in the lexicon of gestalt perception. In their umbral outlines of more complex three-dimensional forms, silhouettes reveal a contest between shape and shapelessness, distilled to its most essential. Drawing silhouettes offered, as the art historian Nancy Forgione has pointed out, "an alternative to modeling" in the sense that provided simpler proxies for volumetric forms.[14] The silhouettes of human subjects were a means of early portraiture thought to fundamentally reveal characteristics of the soul. More recently, the economy of the silhouette has proven a boon to perceptual analysis. For instance, there have been rigorous attempts to quantify the combinatorial range of possible silhouettes against the range of human-recognizable objects.[15] In light of recent research suggesting human object classification is related to the skeletonization of forms, the silhouette seems to enjoy a deep connection to the neurology of human perception.[16]

Certain Measures, the design office I cofounded with Tobias Nolte, has developed a series of projects, two of which we will discuss here, that took the morphological analytics of the silhouette as the basis for reading elements of architectural form and mapping or configuring them in fresh ways. These elements ranged from a fragment of construction waste to the footprint of the building itself. At the heart of these projects was an arsenal of more than 40 different shape metrics that teased

[13] Manovich (2020).

[14] Forgione (1999).

[15] Desolneux et al. (2007).

[16] Ayzenberg and Lourenco (2019).

out the similarity or dissimilarity among the primitive forms of silhouettes. Drawn from the overlapping research of computational morphology, optical perception, and machine vision, these metrics proved strong discriminants among the difference and similarities of silhouette shapes. Our metrics included a suite of shape invariants such as Hu Invariants,[17] shape search and comparison operators such as shape context,[18] spectral comparisons of the graphs of medial axes and polygon skeletons, as well as more simple functions such as normalized perimeter to area ratios and certain functions of our own devising. Using this weighted metric, we induced a dense graph of comparisons among the original set of shapes that reflected their degree of formal similarity or difference. Proximity in this 40-dimensional hyperspace represented gestalt perceptual affinity and formal similarity.

Our dense hyperdimensional network of shape similarity could then be unfolded and flattened into intricate form maps, just as we have unfolded the skins of surfaces. The result maps also become search trees that quantify how easily one might interchange one form with another. In this way, we developed an indirect metric for how unusual a particular form might be. Broadly related to minimum-distortion methods of flattening mesh surfaces, these maps document the hyperspace of shape similarity.

From this map, we can reconstruct a "family tree" of any specific form in the map by determining the other forms which are closest to it and recursively repeating the process. These forms are thus organized into successive generations of similarity. Such a family tree echoes the Bannister Fletcher's famous family trees of architecture or the phylogenetic trees of evolutionary biology.[19] More operationally, a tree of proximate similarity also becomes a search tree to map and associate silhouettes. With it, we can match our design intuition to the closest possible elements with a formal dataset.

8.5 Temporary States of Matter

The computational machinery of silhouette taxonomization can be readily applied to any set of data with polygonal boundaries, from building footprints to room shapes or even outlines of material fragments. It provides enormous versatility and is equally useful as an analytic and form-making device. This mapping and search technique can be combined with methods for the rationalization of complex geometric forms to produce entirely new inversive methods of design that negotiate design possibilities between intended form and a given irregular element dataset.

One example of this process is a first project, Mine the Scrap, a data-driven process that designs new structures algorithmically by reading the shapes of existing scrap (Fig. 8.3). Using we address the pressing need to convert waste into resource. The project scans, classifies, and reconfigures irregular, non-uniform stocks of

[17] Hu (1962).

[18] Belongie and Malik (2000).

[19] Fletcher (1975).

Fig. 8.3 Mine the Scrap, a project developed to leverage machine vision in order to scan waste material and reassemble it into new structures. Certain Measures, 2016

construction scrap into new assemblages and forms, using pattern recognition to find beauty and intricacy in neglected waste.

In this computational process, our 40-dimensional similarity metric is applied to the silhouettes of scrap pieces, that are then reformatted as a series of underdetermined puzzle pieces. Our bespoke software assembles them, squaring irregular parts with a desired whole. The contraction of dimensions attendant to the projection of silhouettes is reversed, and flat silhouette is dilated into spatial volume. The logic at work is a negotiation between what we have and what we desire, a convergence toward a form that satisfies, as best as possible, distinct and irreconcilable demands. It is a design process of exact imperfection. By combining the logic of the quilt with customized shape and pattern detection used in self-driving cars and face-recognition, Mine the Scrap uses big data to tackle big waste. In effect, it creates a search engine for waste derived from machine reading scrap, and then uses this search engine to find the best material solution to design problems.

Mine the Scrap not only creates minimum-waste material lifecycles realized as geometric assemblies, it also develops a new vocabulary of design that is fundamentally informed by the both the horizon of resources and the logics of morphology (Fig. 8.4).

The same techniques for reading, classifying, and mapping forms could be scaled up to the silhouettes and footprints of buildings themselves. To the extent that building footprints inscribe overlapping logics of urban form, parcel geometry, and the characteristic demands of building typology, they comprise an evolutionary intersection

Fig. 8.4 A Machine View of London, a large-scale projection that shows the scanning and morphological organization of one million buildings in central London. Certain Measures, 2019

between extrinsic contextual demands of the city and intrinsic design demands of architectural typology. We have applied this method of reading the urban fabric to dozens of cities and presented the dynamically generated map results in *A Machine view of the City*, a second project consisting of a series of projected video installations that document the process of machine reading the city by form. The resulting maps constitute a comparative metamorphology of architecture aggregated to the city scale. At this aggregated and reconfigured scale, one begins to recognize gradient effects that are a consequence of dense shapes grouping with dense and sparse shapes grouping with sparse. That is, a quality that we might intrinsically identify with parametric variation is in fact a natural consequence of the contextual demands for variation in architecture itself.

The maps generated by machine reading the city also are active, animated documents of the process of machine perception, allowing the user to step inside that process and understand its computational logic. By programming vision-enabled software to scan, read, and synthesize billions of figure-ground shapes and building plans, we can apply data-science techniques to make explicit the formal associations and affinities across the entire corpus of existing buildings. In place of the perennial ad hoc search for meager precedent associations, we could systematically and objectively classify the universe of existing architectural form into a kind of phylogenic tree of shape.[20] In a way, we are returning to a nineteenth-century question of type

[20] Witt (2016).

Fig. 8.5 A Machine View of Boston, a large-scale print of thousands of building footprints classified by form. Certain Measures, 2019

in architecture. The architect Jean-Nicolas-Louis Durand was obsessed with developing a kind of a priori exhaustive catalog or repository of the types of architectural form. At the same time, we were also drawn to the various types of evolutionary trees—not only of organisms, but of physical equipment. We aspire to these kinds of catalogs, but through an empirical ethnography of building, not a priori assumptions (Fig. 8.5).

8.6 Figuring Out Form

Acts of architectural creation might crudely be placed on a spectrum between the poles of figuration on the one hand and configuration on the other. Each mode of creation has a distinct approach vis-à-vis the parts to whole relationship. Figuration, perhaps the more usual and intuitive of the two, evokes a top-down process in which form-making is the primary activity of the designer, and her task is to search for the overall global shapes that best serve or embody a particular architectural intent. Figuration implies a hierarchy in which an overall shape is dominant, and the various parts or components used to realize it are strictly secondary. The figure is the whole in which the parts are arranged.

Configuration inverts the parts-whole relationship of figuration, piecing together a bottom-up patchwork whole from disparate fragments, situating them in a frame or context that lends them a common coherence. The whole thus emerges as a coordination of many figures rather than as the governing gesture of one regulating figure. Configuration arranges elements in a synthetic whole. Configuration is both a mode of design and a mode of mapping, and often one is an essential part of the other. Configuration is also a process well-suited to teasing form from large dataset. It demands the reading and relating of forms to each other in new ways. Machine reading and configurative mapping allow us to externalize and scale up our tacit intuitions across the new media of spatial data.

New modes of machine-readable architecture fuse the practices of figuration and configuration, and mark a new kind of creative process that actively negotiates architectural desires with repositories and datasets of elemental particulars. Machine reading as a generative process strives for coherent form, but through the mechanism of a syntactically rigorous arrangement of particulars. By making the process of relation and association visible, machine-readable architecture blurs the distinction between reading and drawing, perceiving and making, calculating and imagining.

References

Alexander ZÇ (2020) Scanning. In: Alexander ZÇ, May J (eds) Design technics: archaeologies of architectural practice. University of Minnesota Press, Minneapolis
Ayzenberg V, Lourenco SF (2019) Skeletal descriptions of shape provide unique perceptual information for object recognition. Sci Rep 9:9359
Belongie S, Malik J (2000) Matching with shape contexts. In: 2000 Proceedings workshop on content-based access of image and video libraries, pp 20–26. Mori G, Belongie S, Malik J (2005) Efficient shape matching using shape contexts. IEEE Trans Pattern Anal Mach Intell 27(11)
Desolneux A, Moisan L, Morel J-M (2007) From Gestalt theory to image analysis: a probabilistic approach. Springer, Berlin
Dimond TL (1963) Machine reading of handwritten characters. Patent US3108254A
Fletcher B (1975) A history of architecture. The Athlone Press, London
Forgione N (1999) "The shadow only": shadow and silhouette in late nineteenth-century Paris. Art Bull 81(3):497
Forty A (2004) A vocabulary of modern architecture. Thames & Hudson, London, p 72
Gandelsonas M, Morton D (1972a) On reading architecture. Prog Arch 69
Gandelsonas M, Morton D (1972b) On reading architecture. Prog Arch 78
Hu M-K (1962) Visual pattern recognition by moment invariants. IRE Trans Inf Theory 8(2):179–187
Luhn HP (1958a) The automatic creation of literature abstracts. IBM J 159
Luhn HP (1958b) The automatic creation of literature abstracts. IBM J 160
Manovich L (2020) Cultural analytics. MIT Press, Cambridge
Mitrovic B (2009) Architectural formalism and the demise of the linguistic turn. Log 17:17–25 (Fall 2009)
Talbot SA, Harrison WK (1966) Computer evaluation of graphical physiologic data for diagnosis of coronary heart disease. Methods Inf Med 5(2):81–85
Taylor R (1960) Major problems in the use of computing machines. IRE Trans Med Electron 253
Witt A (2016) Cartogrammic metamorphologies; or, enter the Rowebot. Log 36 (Winter 2016)

Chapter 9
Where is Reality? Can You Show It to Me? Constructing Artificial Agency

Theodore Spyropoulos

9.1 Introduction

On 29 March 1971, Heinz von Foerster, physicist and theorist of radical constructivism gave a keynote address at the opening of the Twenty-fourth Annual Conference on World Affairs at the University of Colorado where he stated that "if we don't act ourselves, we shall be acted upon. Thus, if we wish to be subjects, rather than objects, what we see now, that is, our perception, must be foresight rather than hindsight.[1]" For von Foerster no objective reality exists independent of the observer. All things observed are observed by an observer. Observation itself is an act of agency. This is understood as an engaged participation rather than passive witness in constructing a reading of the world. Everyone's understanding of the world is their own, an invention. With this assertion comes a paradoxical underpinning, if everyone's understanding of the world is their own, then this remains inaccessible to others. He confronts us with a provocation that there can be no truly objective world. "'Objectivity is the subject's delusion that observing can be done without him.' The certainty of all forms and their representations therefore are cast into doubt. The only certainty remains uncertainty."[2] Second-order cybernetics defined this understanding by constructing a meaningful break with traditional sciences, the observer here was not outside but inside of the system of the observed. With this individuated agency, facilitating communication is primary. Through communication we may exhibit intelligence, within the context of this paper let us consider this a form

[1] Von Foerster, H. (2003). Perception of the Future and the Future of Perception. In: Understanding Understanding. Springer, New York, NY.

[2] Spyropoulos, T. (2021), Everything You See is Yours: Step Towards the Certainty of Uncertainty. Archit. Design, 91: 64–73.

T. Spyropoulos (✉)
AADRL, Architectural Association School of Architecture, London, UK
e-mail: Spyropoulos_Th@aaschool.ac.uk

P. Morel and H. Bier (eds.), *Disruptive Technologies: The Convergence of New Paradigms in Architecture*, Springer Series in Adaptive Environments,
https://doi.org/10.1007/978-3-031-14160-7_9

Fig. 9.1 Architectural association design research laboratory, London. Working session with students

of architectural or collective intelligence. Intelligence here is not attributed to things as a property but something arising in-between, a product of interface and interaction (Fig. 9.1).

As Director of the Biological Computer Laboratory (BCL) at the University of Illinois Urbana-Champaign, von Foerster and his researchers were invested in the merging of digital and biological systems. Fast forward sixty years the role of the computational within human and non-human systems remains a complex and ever shifting problem. Coupled with the dynamics of our every changing social and ecological landscape our contemporary condition remains paradoxically unexamined. Heinz von Foerster speaks to this conceptual problem with respect to language and experience. He challenges language and its limited ability to communicate. He states, "We seem to be brought up in a world seen through descriptions by others rather than through our own perceptions. This has the consequence that instead of using language as a tool with which to express thoughts and experience, we accept language as a tool that determines our thoughts and experience."[3] The conceptual artist Joseph Kosuth in the creation of his seminal work *One and Three Chairs* in 1965 expands the articulation of the real and the how we understand linguistic and experiential nature through codified registers. A photograph, dictionary definition and object are placed in sequence, each and all fulfilling the descriptive capacity of a chair. The definition, representation and object all respectively function and yet the

[3] Von Foerster, H. (2003). Perception of the Future and the Future of Perception. In: Understanding Understanding. Springer, New York, NY.

assembly of these registers together engages in a thought experiment in understanding meaningful differences. If von Foerster was exploring understanding understanding, Kosuth examined what makes art. If we take Kosuth's chair problem to an extreme conclusion, we could expand this conceptual problem today through a work such as Raffaello D'Andrea's *Robotic Chair* developed in 2006. The chair was designed to disassemble and reassemble itself. Each elemental component of the chair, its legs, seat and backrest all understood as its constituent parts in the reassembly of itself. Rather than an object, this chair exhibited agency and goal orientation that searched for its unitary parts to fulfil its assembled goal. We can consider objects today as seeing and being in their respective worlds. In this work degrees of chair-ness are registers that point to a world of things and their presences in world building. A world of worlds that makes no distinctions between human and non-human things. "We live in a technologically interconnected social and ecological sphere that has made us hyperaware of the magnitude and complexities of the challenges today. By necessity, if contemporary design is to remain relevant it must shift from the finite representational models of practice towards real-time collaborative ones. The shift conceptually is to move from 'models of' towards 'models for'."[4] In framing these emerging challenges, it is important to situate architecture as an adaptive framework that can offer an alternative model of how we understand and situate things in the world. Architecture here is adaptive and behavioural. It moves beyond the fixed and finite understanding towards a behavioural model for architecture that is adaptive and evolving (Fig. 9.2).

A behavioural architecture challenges blueprints and master plans and articulates real-time models that evolve, learn and adapt through time and their capacity to be self-aware and self-structuring. The research featured looks to systemic features articulated through abstraction in an attempt to generalize higher population of interacting agents. This process of digital breeding and competition-based environments can be clearly illustrated through the seminal work of Karl Sims in his papers on the subject written in the mid-nineties such as Evolving 3D Morphology and Behavior by Competition.[5] Sims writes that "In natural evolutionary systems the measure of fitness is not constant: the reproducibility of an organism depends on many environmental factors including other evolving organisms, and is continuously in flux. Competition between organisms is thought to play a significant role in preventing static fitness landscapes and sustaining evolutionary change." Rather than privileging prescriptive models, the genetic pool evolves and tests relational and population dependent organizations that aim to perform through locomotion. This process affords a design plurality of plausible solutions, performing as a body or creature for duration of time before other higher order goals are learned. The aim of this process is to evolve creatures that have the capacity to have self-awareness and autonomy of

---

[4] Spyropoulos, T. (2021), Everything You See is Yours: Step Towards the Certainty of Uncertainty. Archit. Design, 91: 64–73.

[5] Sims Karl (1994), Evolving 3D Morphology and Behavior by Competition, Artificial Life (Volume: 1, Issue: 4, July 1994).

Fig. 9.2 Minimaforms (Stephen and Theodore Spyropoulos). Project: memory cloud, Detroit institute of arts (2012). *Photo* Theodore Spyropoulos. Description: Memory Cloud is a participatory framework based on the ancient practice of smoke signals—one of the oldest forms of visual communication. Fusing ancient and contemporary mediums, Memory Cloud creates a dynamic hybrid space that communicates personal statements as part of an evolving text, animating the built environment through conversation

control to allow each organization to have local and global awareness. This evolutionary model for design examines how high populations of units could interact and through this interaction develop features that could evolve the system to be self-aware, self-structure and assemble. Goals such as mobility and self-structuring are the main drivers for this research as it stands today. Environmental conditioning, machine learning and collective building expand territories of communication that speculate real-time interaction of space as a continuous dynamic system of formations enabling a framework to acknowledge the evolutions of machines capacity to have meaningful interactions with other machines. Speculations of this form of engagement date back to mathematician John von Neumann's thought experiments in the late forties on a kinematic model for a physical self-replicating machine to more contemporary research by Nissan with their self-parking robotic office chairs. Enabled through programmable matter, actuated soft robotics and embedded sensing technologies behavioural complexity offers new terms of reference for architecture. Architecture of the future present will engage us, challenge us and enable a new species and taxonomies of proto cybernetic ecologies.

"English psychiatrist William Ross Ashby in his landmark book titled *An Introduction to Cybernetics* published in 1956 articulates its early conceptual framework when he states, "Cybernetics… is a "theory of machines" but treats, not things but ways of behaving. It does not ask "what is this thing?" but "what does it do?" … It takes as its subject matter the domain of "all possible machines," and is only secondarily interested if informed that some of them have not yet been made, either by Man or by Nature. What cybernetics offers is the framework on which all individual machines may be ordered, related, and understood."[6] Behaviour as subject in early cybernetic discourse made little to no distinction between objects, organisms or machines, and only considered agency as a product of an entities capacity to produce change in an environment. This served as a fundamental driver for the behavioural classification proposed in the seminal paper titled "Behavior, Purpose, and Teleology,"[7] published in 1943 by authors Arthur Rosenblueth, Norbet Wiener, and Julian Bigelow which influenced some of the core conversations at the cybernetics conferences held between 1946 and 1953 at the Josiah Macy, Jr. Foundation.[8] Furthering Ashby's questioning of what things do, Andrew Pickering in his book The Mangle of Practice: Time, Agency and Science makes an important distinction with what he sees in second-order cybernetics as shift from "the representational idiom" to what he states as "the performative idiom." The representational idiom maps the world and describes it as it is, while the performative idiom is concerned with agency and influencing this world through action (Fig. 9.3). Pickering sees this as "the emergent interplay of human and material agency".[9] Within the context of this paper, it is the "interplay between human and non-human agency."[10]

Second Order Cybernetician Ranulph Glanville expresses a behavioural understanding when stating, "while we all observe and know differently, we behave as if we were observing the same thing. What structure might support this? One supporting the essential difference while retaining the possibility of communication: when the basic assumption is that we are all different, we all see and understand differently."[11] Communication and the interface of our interaction with each other, our environment or with non-human agents therefore cannot be assumed. Sociologist Andrew Pickering argues that characterized by graspable causes,

---

[6] Johnston, John, 'The Allure of Machinic Life: Cybernetics, Artificial Life, and the New AI', The MIT Press, 2008. Pg.11.

[7] Rosenblueth, Arturo; Wiener, Norbert; Bigelow, Julian (Jan 1943). "Behavior, Purpose and Teleology". Philosophy of Science. 10 (1): 21.

[8] The Macy Conferences brought together a diverse group of cross-disciplinary scholars that included mathematician and computing pioneer John von Neumann, founder of cybernetics Norbert Wiener, social scientist Gregory Bateson, cultural anthropologist Margaret Mead, biophysicist Heinz von Foerster, father of information theory Claude Shannon, amongst others. The meetings were foundational in the development of cybernetics and systems theory.

[9] Pickering, Andrew (1995). "The Mangle of Practice Time, Agency, and Science." Chicago, Illinois: University of Chicago Press.

[10] Spyropoulos,T. (2017). 'Constructing Participatory Environments: A Behavioural Model for Design', PhD dissertation, University College London (UCL), 2017, p 24: https://discovery.ucl.ac.uk/id/eprint/1574512/.

[11] Glanville, R. (2003) 'Second-Order Cybernetics', EoLSS Publishers. Available at: http://www.eolss.net/sample-chapters/c02/e6-46-03-03.pdf.

Fig. 9.3 Minimaforms (Theodore and Stephen Spyropoulos), of and in the World, London, 2017-present. Description: Fifteen hundred glass orbs construct an inhabitable crystalline lattice that examines world modelling by exploiting physics; optics and our dynamic readings of space through our experience. The orbs use cosmological and celestial organizations of three offset spherical layers that interconnect each of the orbs in a dynamic equilibrium. The work in conception and realization is a construct of the mind, demonstrating what one sees is truly their own

but rather of one in which reality is always "in the making," to borrow a phrase from William James. We could say, then that the ontology of cybernetics was non-modern in two ways: in its refusal of a dualist split between people and things, and in an evolutionary, rather than casual and calculable grasp of temporal process."[12] The situated complexity of observers, the environment of this observation, and the potential to draw out communication and shared experience motivates the underlying premise argued as a behavioural framework for design. The role of this framework to engage in real-time with the complexities of communication within a collective is hypothesized through a model for interaction as conversation. Enter design.[13]

[12] Pickering, A. (2010) The Cybernetic Brain: Sketches Of Another Future. Chicago, Ill: University of Chicago Press, pp. 309–371.

[13] Spyropoulos, T. (2017) 'Constructing Participatory Environments: A Behavioural Model for Design', PhD dissertation, University College London (UCL), p 28: https://discovery.ucl.ac.uk/id/eprint/1574512/.

The world within worlds is a radical constructivist approach that foregrounds agency and the desire to express communication and the primary demonstration of intelligence and understanding. Architecture here is understood as something relational and plural responding to a world that is latent and unknown. Agency here is seen as a goal-seeking attribute that speaks to this evolving relationship over time.

Marshall McLuhan would suggest that what we consider the future is actually our present because we find ourselves living in the past. Our contemporary age is as radical with change, latency and uncertainty becoming the new norm. Considering our present, we may agree we live in an age where science fiction has become fact. The need for architecture to engage socially and participate in the challenges of our time is fundamental. Architecture, academia, and the construction industry today remain overtly conservative, habitual and singular in their approach. The pursuit of exploratory knowledge and the plurality of design problems will be the only manner in which we may make a meaningful impact. From mass migration to climate change, architecture in its conception and practice can no longer be considered something fixed and finite. We must understand deeply our world as one that necessitates systems and thinking that is adaptive and evolving. We must also consider the orthodoxy of styles and historical crutches as limited if not obsolete in this pursuit. Our built environment should enable more participatory means to share and explore space as a medium of our interfacing. Technology in this pursuit is our enabling framework to bring us together to rise to these challenges (Fig. 9.4).

Fig. 9.4 Minimaforms (Theodore and Stephen Spyropoulos), the order of time, Le Quadrilatère, Beauvais, 2022. Exhibition "Yona Friedman L'EXPOSITION MOBILE in dialogue with Minimaforms". *Photo* Salim Santa Lucia. Description: The Order of Time is an attempt to consider our environments as universal, mobile, and ordered through spatial relations. Our installation builds on Friedman's ideas of space-time to an extreme conclusion. A space constructed as an evolutionary computational architecture. Examining the attempts to see mathematics, science and matter as facilitators to this open and evolving framework

Chapter 10
From Disruptions in Architectural Pedagogy to Disruptive Pedagogies for Architecture

Sevgi Türkkan

10.1 Introduction

Only two decades into the twenty-first century, the disciplinary tapestry of architectural education, woven by deeply engrained values, institutional structures, long-resistant cultural practices, and shape-shifting yet continuous traditions is already confronting an unprecedented set of disruptions.

It is not the exponentially growing range, capacity, and precision of computational tools at the service of spatial design, production, and operation, alone; nor the advanced communications between human and non-human agents, data, and expertise across time and space; nor artificial intelligence's striking abilities to learn, imagine and even dream. It is also the drastic urgencies of the climate crisis, social and environmental inequalities, added to the globally experienced despatialization[1] of architecture schools during the COVID-19 pandemic that altogether reveals the obsolescence of the hard-wired, long-standing conventions of architectural education.

In the past 20 years of architectural education, computers ceased to be a technology either to be celebrated or resisted, but simply a fact of life (Allen 2012). CAD/CAM infrastructure and ICT platforms have become ubiquitous for design learning environments in most architecture schools in the world. An increasing number of master and PhD programs, research clusters, and design studios in institutions mostly stemming from the polytechnic tradition pursue sophisticated, innovative design research projects on computation and fabrication in spatial practices. Moreover, laboratories, such as the MIT Media Lab, push the boundaries of digital technologies and design knowledge beyond the territories of traditional disciplines.

---

[1] As coined by W. Mitchell in City of Bits in 1996.

S. Türkkan (✉)
Istanbul Technical University, Istanbul, Turkey
e-mail: turkkan@itu.edu.tr; sevgiturkkan@gmail.com

P. Morel and H. Bier (eds.), *Disruptive Technologies: The Convergence of New Paradigms in Architecture*, Springer Series in Adaptive Environments,
https://doi.org/10.1007/978-3-031-14160-7_10

However, although new digital design concepts and practices have led to reconsiderations for digital design pedagogy,[2] it is hard to claim that the overall cultural establishment, curricular structure, organization model, or value system of mainstream architectural education has been drastically interrupted by the first digital turn. The cultural practices, motives, codes, and rituals that mostly stem from the nineteenth and twentieth century institutionalization period are still largely in place in cultivating the architect figure and his[3] disciplinary territory. In schools around the world, jury reviews continue to be held via high-resolution, dynamic, computer-generated graphics, nowadays via computer screens. Competition and charette culture thrive over social media platforms and networks. The curricular disinterest toward labor, cost, impact, or feedback processes continues to isolate the design studio from everyday actualities. Relying on satellite images, street views, and data sets provided by Google and others, projects can be executed anywhere, engaging with a distant site or context from the comfort of our studios (now homes). Buildings are still regarded as the superior form of spatial practice, while their temporal nature is mostly disregarded. The quest for originality and novelty continues to drive form-finding efforts in changing disguise. A project's visual and rhetoric power is still the primary gadget to claim students' authorship, therefore praise, acknowledgement, and grade individually received.

While certain institutions undertake cutting-edge explorations in the field, and every architecture school is being inevitably immersed into digital culture (the pandemic made the final stroke), there are pending disciplinary, cultural, and methodological questions on what new forms of intelligence, labor, creativity, and reorganization of space, knowledge and resources have to offer for architectural learning.

More so, the social and climatic urgencies compel architectural education to radically reconsider a new pedagogic agenda in the age of big data, AI, and the Anthropocene.

This article proposes to outline trajectories for this agenda, by raising a series of questions regarding architectural learning and the role of institutions in the twenty-first century. Some of these questions are new, many date back to the institutional foundations of architectural education that call for a revision as digital culture and the impacts of the Anthropocene advance in unprecedented speed.

The arguments in this essay around architectural pedagogies do not primarily concern the laboratories or programs that pursue cutting-edge experiments on the verges of the discipline. The discussed disruption scenarios address directly the mainstream, ordinary architecture school, its educational concepts, curriculum, pedagogic rituals, values, and the disciplinary ethos that lies underneath it. Certainly, speculations on future spatial practices and education could exceed the limits of our imagination depending on how much we fast-forward in time. But we do not need to jump

[2] Rivka Oxman, Digital architecture as a challenge for design pedagogy: theory, knowledge, models and medium, Design Studies Vol. 29, No. 2, March 2008.

[3] The reader may observe a shift in the use of pronouns regarding the architect-figure throughout the essay. It is intended to play as an indicator of the gender biases of mentioned cultural contexts.

into an unforeseen future for speculation. The state of architectural education today is already puzzled with its unpreparedness for alarming social and environmental crises, overwhelming sophistication of the tools of the "future", and its folkloric rituals and anachronistic conventions.

In order to tackle the disciplinary, epistemological and techno-cultural challenges and disruptions, the essay will launch trajectories defined by the three founding archetypes of architectural education: *the school, the learner,* and *the project.* By resorting to cases, practices, and discourses on architecture, digital culture, and education, it is aimed to discuss and reconsider these disruptions as emancipatory points of departure for a renewed architectural pedagogy. The foundation of this essay lies in the rather uncharted intersection between architectural design, digital knowledge, and critical pedagogy.[4]

By way of speculation, it is not intended to reach prescriptions or convictions, but to uncover the important questions for cultivating a more informed, responsive spatial practice and imagination, possibly through critical digital architectural pedagogies.

10.2 Disruption #1: The Architecture School

From the Centrally Located School to Deschooled Centers

The geographic location of architecture schools has long been a token of its anticipated mode of engagement with the world. It was the weighty, insular establishment of the École des Beaux-Arts on the site of a former convent in the center of Paris, Europe's nineteenth century artistic and cultural capital that helped sustain its traditions and status quo for centuries. Yet, it was the same centrality that led to its demise during the May 68 student revolts, which resulted in its dispersion into several *Unité Pédagogiques* across Paris.

The 1960s saw the flourishing of ideas on deschooling, decentralization, and despatialization of educational institutions, due to expanding global mobility, telecommunication infrastructure and disruptive scholarship such as critical pedagogy. Illich's Deschooling Society (1971) famously advocated the disestablishment of schooling for its role in perpetrating an unjust social order and called for radical new ways of learning. He proposed entangled educational networks called "learning webs"; communication infrastructures that function outside the remit of the school, combining educational objects, peer learning, mentorship, and reference services.[5]

[4] There is increasing scholarly work bridging data science and the field of education (critical digital pedagogy, digital literacy, data feminism), between computation and architectural education, as well as critical pedagogy and architectural education. Yet the intersection of digital knowledge, critical pedagogy and architectural education is still a rather unexplored area.

[5] https://www.e-flux.com/architecture/education/322673/deschooling-architecture/.

His idea was to create a framework that "constantly educates to action, participation, and self-help."[6] De-institutionalizing education was connected to a concern for conviviality; the ordering of education, work, and a society in line with human needs, therefore to 'de-professionalization' of social relations.[7]

Other critical pedagogy scholars pointed out the role of situatedness in learning. Freire hinted at the spatial aspect of "situationality": "People, by being 'in a situation' find themselves rooted in temporal-spatial conditions which mark them and which they also mark. They will tend to reflect on their own 'situationality' to the extent that they are challenged by it to act upon it."[8] Gruenewald (2003) synthesized the discourses of place-based education and critical pedagogy into a "critical pedagogy of place" with an emphasis on ecological thinking that social analysis often disregards. The "critical pedagogy of place" connects to a wide range of learning models,[9] and offers an agenda of cultural decolonization and ecological reinhabitation.

Similar thoughts provoked ideas on the spatial reestablishment of architecture schools: Giancarlo De Carlo's radical proposal for a decentralized university (1962–65), the mobile network of academic structures designed by Cedric Price in Potteries Thinkbelt (1965), Candilis, Josic, Woods' open-system building for the FU in Berlin (1967–73),[10] the TV broadcast "A305, History of Architecture in Design"[11] in The Open University (1967) can be counted among the influential thought-experiments that geographically reconfigured architectural learning. In the same period, those who could not dismantle their schools left their buildings to invent new engagements with the world outside.[12]

The introduction of early digital technologies opened a new window of opportunities for designing in multiple space and time, and inspired architectural experiments in a pedagogic context: Frazer's Universal Constructor at the AA London, Bits and Spaces group in ETH Zurich, Paperless studios in GSAPP New York, launched series of pedagogic experiments on the capacities of virtual space using early CAAD software and dial-up Internet, exploring co-located multi-user interaction, distant collaboration, open-source and 3d models. In the 2010s computation-based programs advanced their distant design operations with robotic automation, BIM, and other

---

[6] Ivan Illich, *Deschooling Society* (London: Marion Boyars Publishers Ltd., 1995), 32.

[7] https://infed.org/mobi/ivan-illich-deschooling-conviviality-and-lifelong-learning/.

[8] Freire (1971) continues, "Human beings are because they are in a situation. And they will be more the more they not only critically reflect upon their existence but critically act upon it".

[9] Gruenawald (2003) exemplifies the uses of PBL in "experiential learning, contextual learning, problem-based learning, outdoor education, indigenous knowledge, environmental and ecological education, bioregional education, democratic education, multicultural education and community-based education, critical pedagogy, as well as other approaches that are concerned with context and the value of learning from and nurturing specific places and communities.".

[10] https://www.architectural-review.com/today/radical-pedagogies-in-architectural-education, retrieved 07.03.2020.

[11] https://www.cca.qc.ca/en/events/50959/the-university-is-now-on-air-broadcasting-modern-architecture, retrieved 07.03.2020.

[12] Some well-known design studio practices: Learning from Las Vegas, Rural Studio, Open City in Valparaiso.

hybrid collaborative environments for co-design and manufacturing. Also, workshops, summer schools, and satellite laboratories stretched overseas, although some could not survive the logistic challenges.[13]

What the pandemic made clear was that the relevancy of an architecture school depended on its capacity to establish meaningful engagements with various localities (including its own) and reflect on its geopolitical agency. This concern raises questions both on the level of the school's self-organization (spaces, knowledge, and teaching resources), as well as the modus operandi of the design studio. After all, the problematic centrality of the studio tutor in the design studio as the primary and dominant source of knowledge can be seen as an epitome of the school's relationship to the world.

How does the architecture school today engage with the "outside" world in its everyday design-learning environments? How does it acknowledge the geographical scale it operates in? How does local knowledge (students and projects dispersed around the world) enter the learning process, and how are resources shared with the outside? How does locality play in situating the content and process of design learning in the age of information? What does the despatialization of architectural learning and education mean in the geopolitical context?

As the pandemic exposed, the pressing social and climate inequalities amplify the significance of these questions, considering the impacts or contributions of spatial practices on built, social and natural environments. Typically located in privileged campus sites or urban capitals, architecture schools are confronting the need to be more responsive and receptive in their agendas, strategies, and tools, and find new ways to engage with cultures and knowledge across the globe. The emergency online solutions may have facilitated a swift continuation of architectural education at a global scale. But disconnection and isolation historically embedded in the cultural practices[14] of the design studio are still waiting to be structurally addressed as a pedagogic, curricular, and disciplinary problem.

In the past decade, there were significant developments in distant learning (online learning,[15] digital classrooms, learning apps,[16] smart campuses, MOOC's, adaptive learning environments[17]) and distant spatial practices (responsive, adaptive systems, automated, immersive robotic environments, digital twins). Surveillance technologies, remote sensing, satellite imaging, open-source and cloud-based platforms enabled information to be produced, shared, and communicated by agents

---

[13] Such as Columbia University's Studio-X program.

[14] Largely stemming from the competitive nature of design studios and formulation of design process as composition.

[15] Architecture Education is Unhealthy, Expensive, and Ineffective. Could Online Learning Change That? https://www.archdaily.com/884590/architecture-education-is-unhealthy-expensive-and-ineffective-could-online-learning-change-that?ad_medium=widget&ad_name=navigation-prev.

[16] Tweeting from the Tower: Exploring the Role of Critical Educators in the Digital Age, Anderson, Morgan; Keehn, Gabriel, Critical Questions in Education, V. 10, N. 2, pp. 135–149, 2019.

[17] Santoianni, Flavia & Ciasullo, Alessandro (2018). Adaptive Design for Educational Hypermedia Environments and Bio-Educational Adaptive Design for 3D Virtual Learning Environments. Research on Education and Media. 10. 30–41. 10.1515/rem-2018–0005.

in various locations. Considering the geopolitical urgencies of climatic, and social issues, and given the affordances of new digital infrastructure, the architecture school's engagements with place (operations, organization and pedagogic agenda) can be rethought beyond the conventional methods of site analysis, excursions, and one-way interventions, or the latest Zoom-enabled formats of online learning.

Can recent advancements in digital knowledge, big data and AI enhance architectural learning to be more critically engaged with place? How can the convergence of physical and digital design environments contribute to building long-term, multi-layered, real-time conversations with localized agencies in the design studio? What are the affordances of adaptive, responsive, immersive, networked environments and immersive media (VR/AR/MR) to enhance embodied presence for place-based learning? Can computation (data analytics, machine learning, sensory networks, automation) help extract complex layers of situated knowledge, intelligence, resources that are material, immaterial, immeasurable and temporary? Can design-learning processes be reorganized beyond the capabilities of studio tutors, to accommodate bio-cyber-virtual-physical feedbacks from human and non-human agencies, intelligences and expertise? Can spatial learning be made more accessible ? (Fig. 10.1).

A critical transformation in the design-learning and schools' geopolitical role seems to lie in the technological and conceptual reinvention of the "vernacular" in

Fig. 10.1 Le Grand Salle interior façade of the École des Beaux Arts, ETH Zurich "Concrete Choreography", Forensic Architecture's 3D Model reconstruction of the "Black Friday" 2014, railways from the Potteries Thinkbelt, an anonymous drone, Children in Negroponte's One laptop per child project, The guardian of the loge during an architectural competition (see image 2). Collage executed by Anıl Aydınoğlu

the new digital turn. As opposed to one-way, top-down imposition of expert knowledge from the "ivory tower",[18] computation can be employed to discover new ways of making sense of the vernacular: urban or rural complexities, intelligences, material and immaterial resources, socio-economic patterns, labor and craft knowledge, biodiversity, etc. (i.e., Feral Atlas[19] and Forensic Architecture are two inspirational practices that use digital techniques[20] resourcefully in investigating the spatio-temporal complexities of lived events and environments). The digital infrastructure itself is a pedagogic inquiry into new forms of vernacular spatial knowledge. Such a techno-cultural-pedagogic shift would nourish the design studio practices to embrace sensitive, generous and adaptive approaches to place beyond the habits of relying on satellite images, found information, and generic assumptions. Place-based pedagogies and adaptive environments for design learning could turn the knowledge-production processes into real-time, multi-directional, long-term dialogues with situated agencies. With digitally informed critical pedagogies, the role of the school could ideally shift from delivering the western canon to co-producing geopolitically relevant and site-specific spatial knowledge.

The idea of moving from the cathedral to the bazaar[21] may be too ambitious considering schools' heavy institutional bodies, bound to national policies, regulations, and market expectations. Yet, the need to act upon local and global urgencies compels the schools to rethink the relevancy of their operations at every level.

A 'technologically reinforced critical pedagogy of place' calls upon the creativity and resourcefulness of school administrators, curriculum designers, teaching staff, as well as researchers and producers from fields not limited to architecture. Considering the complicity of architecture schools in canonizing western history, production modes, and anthropocentric world-view, a revision of situated spatial learning promises exciting opportunities for decolonizing architectural education, undoing the damages caused by local insensitivities, and more flexibility in providing spatial responses to global social or climatic catastrophes.

[18] How critical pedagogy scholars commonly refer to the nineteenth century establishment of educational institutions.

[19] https://feralatlas.org/, retrieved 21.03.2020.

[20] Forensic Architecture uses 3d modeling, audio analysis, data mining, field work, fluid dynamics, geolocation, ground truth, image complex, software development, machine learning, osint, pattern analysis, photogrammetry reenactment, remote sensing, shadow analysis, situated testimony, synchronization, virtual reality (https://forensic-architecture.org/about/agency).

[21] "The Cathedral and the Bazaar" is the title of Eric S. Raymond's book (1999) that uses the spatial analogy to discuss the development of Linux-based operating systems and the open source systems as new dynamics of information economy.

10.3 Disruption #2: The Architecture Learner. From Author to Agent to Cyborg

Architecture schools are the reproduction mechanisms of the discipline. They function by transmitting the "symbolic capital", the repertoire of culture and skills needed in order to reproduce itself (Stevens, 2002). Schools essentially contribute to this mechanism by "authorizing" architects. This happens mainly in two ways: shaping and providing the formal[22] educational repertoire prerequisite for professional certification, and secondly, by instilling author-like behaviours, responses, mind-set, and values through the everyday experiences of the pedagogic culture.

The task of "making one into an author" has led architectural schools to develop formal, informal, explicit, and hidden pedagogic practices that nourish and test one's individual skills and capacities in becoming an architect-figure. The École des Beaux-Arts even built a spatial mechanism[23] consisting of individual cells aligned on a corridor, surveilled by guardians, isolating students physically and socially during architectural competitions in durations ranging from 2 h to 3 months. To counter the anonymity in the atelier, this system enabled testing students' authentic design skills and architectural virtuosity (the diligent renderings of their initial sketches) without assistance or interruption, annihilating any doubt that might shadow their individual authorship.

Along the twentieth century,[24] the questioning of the heroic-solo-creator myth inspired creative practices including architecture and its education to explore alternative conceptions of authorship. "Agency" emerged as a theoretical alternative against the predicaments of the Albertian architect-figure in spatial practices (see Spatial Agency[25]). Also, the transition from mechanical to digital technologies, and from identical to variable reproductions, enforced the need to recast "architectural agency" (Carpo 2011).

Computation's vast appropriation of the term "agency" indicates the ambitions for abdicating human's authorial decisions and functions to the machines. Today, artificial intelligence, machine learning, expert systems, and neural networks enable "intelligent agents" (to varying degrees) to possess autonomy, mobility, a symbolic model of reality, a capacity to learn from experience, and an ability to cooperate with other agents and systems.[26]

---

[22] According to criteria agreed upon by national, international, regional boards, governmental policies, professional chambers and institutes.

[23] Competitions, exhibitions, diploma project, jury reviews, desk-crits, credit systems are some of the pervasive and still used pedagogic inventions of École des Beaux-Arts for the individual evaluation of the student-architect.

[24] Notably in changing techno-social paradigms such as 1930s mechanical reproduction, the late 1960s post-structuralist theories and creative cultural practices, 2000s digital culture.

[25] Awan, N.; Schneider, T.; Till, J. (2011). Spatial agency: other ways of doing architecture, New York, NY: Routledge.

[26] https://www.britannica.com/technology/agent.

Further conceptions of the agency came to life due to bidirectional informational frameworks (such as P2P and distributed processing networks, cloud computing)[27] allowing larger numbers and a variety of agents (human, non-human, experts, non-experts) to collaborate in a "social" manner. Digital mass customization (theory of the objectile) has broadened the spectrum of agency by implying a layered model of authorship, or "split agency", where the primary author designs a generic (parametric) object, and one or more secondary authors adjust and adapt some variable aspects of the original notation at will (Carpo 2017).

Moving on to the spatial agent and the hybrid co-agency, the human designer today can operate through an unprecedented spectrum of technologies and intelligences available for her operational, creative, communication, and manufacturing endeavors. Human and non-human agents, enabled by computation or virtual presence, can be assigned as design assistants, partners, and collaborators. The plethora of options in the medium of collaboration, possible organizations between human, bio, cyber actors and intelligences, technological range of production tools, put the architect in a unique position to choose and operate through a radical new set of partnerships and affordances.

Yet, are architects trained for non-conventional types of authorship? How are other agencies pedagogically approached in a design process? Do schools provide the organizational skills and knowledge to choose, assign or inform agencies? What modes of shared/split spatial agencies do students experience in the design studio (besides team-work)? Is the designing of communications, interactions, exchanges, and protocols between the agents considered part of the design process?

The challenge to these questions is presented in the beginning. Architecture schools are structurally designed to authorize individuals through their original authentic creations, reflecting "the will and mind of a (single) 'creator'.[28] Some schools build their reputations on granting individuals the skills for greater authority and autonomy in their professional lives.[29] Even innovative pedagogic experiments like the Columbia Building Intelligence Project (CBIP) that explore new technology-enabled design collaboration and alternative authorship models as "creative options to BIM and IPD" do not shun from asserting the architect's leader role in future industries. The motivation to "expand the scope and capabilities of architects" is articulated for the purpose of "embedding the role of design in the total process of realizing a building".[30]

Architects' historical preoccupation with individual subjectivity and autonomous creative power has long been dubious due to their inextricable dependency on tools and relationship to others in performing architecture (even the most rigid cubicles

---

[27] Carpo, M. (2011). The Alphabet and the Algorithm (p. 113). Cambridge, MA: The MIT Press.

[28] Anstey, T. (2003). Authorship and Authority in L.B. Alberti's De re aedificatoria. *Nordisk Arkitekturforskning,16*(4), 19–25.

[29] At least that is the expectation behind the astronomic tuition fees.

[30] http://www.columbia.edu/cu/arch/courses/syllabi2/A4104/2014/1/006/A4104_006_2014_1_Marble.pdf (p. 13).

Fig. 10.2 Cartpostal with the text "Rennes—Ecole d'Architecture—Un jour de loge" displaying the competitiors and guardians in the loge, 1930, Steampunk Pavilion augmented hands-on construction in TAB 2019, Estonia, Kismet, Alexa, Siri, Architects Declare poster, The Silk Pavilion bio-digital fabrication project, Extinction Rebellion, Zizi the deepfake AI drag persona, ON AIR ecosystem by Saraceno (2019), DeepDream by Google. Collage executed by Anıl Aydınoğlu

of the Beaux-Arts could not prevent that[31]). The intellectual and creative capacities of new agents complicate the boundaries of human body and mind to a level hard to distinguish. Haraway's cyborg as "a hybrid of machine and organism, a creature of social reality as well as a creature of fiction"[32] is too present to be condoned by any educational practice today, including architecture.

So, *who is the new subject of architectural learning? Who is subjected to assessment and authorization* via *architecture pedagogy? What do schools offer to the human designer in the age of deep-computation and data-driven epistemology? How is performance assessed in a data-driven, multi-agent pedagogical experience? Does the question of "who" still matter? Should schools continue looking for an author?* (Fig. 10.2).

The architecture learner has historically been the human, precisely, the white European male. In the age of big data, machine learning and AI, the subject of architectural learning in schools is still primarily the human. After all, it is he who establishes the socio-economic drive toward deep machine learning and who builds the vast computational power to perform it, while it is also the human who ends up

[31] There are numerous archival accounts of student memoirs reporting incidents of cheating, sneaking in ready-made drawings, and using cigarette breaks to discuss and help each other with their projects.

[32] Haraway (1985: 65) cited in Smith & Selfe (2006).

problematizing the immense carbon footprint it produces, and is compelled to find solutions to prevent further contributions to global warming.[33]

Yet it is a new human that architectural pedagogy is dealing with. Consisting of "partial, contradictory, permanently unclosed constructions of personal and collective selves," (p. 75) this new human cannot be addressed with old pedagogies that single out names and aesthetic objects, while leaving aside the relationships, performances and consequences that emerge. There is a need to pedagogically recognize plurality, by creating new cultural habits and rituals, as well as an assessment structure that values and credits agents with their contributions.

There are still reasons to maintain authors and authorship structures in a pedagogic context. After all, they regulate liabilities for the political, socio-cultural, and environmental ramifications of design decisions. Taking from art and software design, alternative legal certification systems[34] and authorship models provide fruitful substitutes for managing the credits and responsibilities (even if complicated by big data and AI, as in the case of the self-driving car accident[35]). Maybe then the focus could shift onto the evaluation of the qualities that make a project or practice "good" (see Disruption #3).

How would this shift affect the schools' responsibility toward the learner? Two types of empowerment seem crucial for individuals to navigate in the age of information and global catastrophes without falling into technological determinism or optimism: (1) Digital literacy, knowledge, and skill-base to operate a complex range of technologies; (2) Critical thinking skills to situate and (mis)use information and software toward a place-based, socially-oriented practice. Besides a curricular readjustment, the school's biggest liability to the learner is to provide a resourceful platform that facilitates and triggers experimentation with agencies, modes of practice, technologies, materialities, and creative spatial imagination.

10.4 Disruption #3: The Architectural Project. Reintroducing Concepts and Concerns

Schools' pedagogic agendas have always played a role in reframing the "good practice" of an era. The Beaux-Arts prized the flawless rendering of a well-composed plan; Bauhaus esteemed discovery of modern life through materials and forms; the Texas Rangers valued contextuality over modernism; Boyarski provoked research

[33] The computing power required for AI landmarks increased 300,000-fold from 2012 to 2018 (https://www.wired.com/story/ai-great-things-burn-planet/).

[34] Creative Commons' licenses such as copyleft, share-alike, etc., also forking in github, appropriation art.

[35] https://www.forbes.com/sites/fernandezelizabeth/2020/02/06/who-is-responsible-in-A-crash-with-A-self-driving-car/?sh=ef210854b2b6, retrieved March 27, 2021.

through architectural representation. Scott Brown and Venturi acclaimed the ordinary as a pedagogic resource. Tschumi's Paperless Studio ventured into the new discursive capacities of notations in the early CAD environment.[36]

As the software systems become more invisible and infrastructural (Cardoso Llach 2015), qualities other than technological advancement, such as methodological sensitivities and positions, come to the fore with new concerns for pedagogical evaluation.

So, what is a "good" architectural (student) project today? What values, competencies or qualities are expected from it? What new values are esteemed in the age of big data? What new qualities of architectural performance could be endorsed with critical digital architectural pedagogy?

Let's discuss a series of qualities that could reframe a "good" project today:

How creative and sensitive is the design of the design process? Does the design of the process align with the project's overall motivations? What designerly qualities appear in the operational choices considering the tools, platforms, software and resources, human–computer interactions, and social organization of the agents? Was the collaborative and intelligent nature of digital infrastructure used creatively and resourcefully?

How responsive is the project toward spatio-temporal specificities? Does the project produce locally informed, situated strategies and outcomes? Are design protocols and interfaces employed to establish dialogue, empower or give voice to local agencies (as discussed in Disruption #1)? Does the project partake in producing vernacular spatial knowledge "as opposed to generic, objective disembodied knowledge" (Harraway 1988)?

Are the temporal dimensions of design, construction and use recognized and addressed in the project? How does temporality (from production to the afterlife) contribute to the design idea and construction process beyond time-saving concepts? How are the new digital affordances (simulations, real-time digital models, automation, etc.) utilized in reorganizing spatio-temporal relations in the service of the design intention? Is the architectural performance conceived and imagined in ranging time-spans to consider alternative labor, cost, and impact scenarios?

Were the data and software tools exercised with rigorous technical knowledge and critical awareness? Today, even the most conventional design studio cannot escape the modus operandi of the information age, due to the ubiquity of search engines, satellite imagery, drawing, and imaging software. May (2017) argues that "our contemporary condition is thus marked by a kind of servile lusting after the 'data products' which architects and urbanists know how to use, but know nothing about."

Architectural education's troubled history with data products is famously marked by the decades-long popularity of the Architects' Data (1936).[37] An "architectural

[36] https://www.cca.qc.ca/en/articles/issues/4/origins-of-the-digital/33488/paperless-studios, retrieved April 2, 2021.

[37] As a 1926 Bauhaus graduate, Neufert observed the paradoxes between his master Gropius' lectures on rationalization and industrialization of production processes, and the inefficiencies in

database for timesaving design process and efficiency-based working methods" (Weckerlin, 2007) resonated with architectural education, as strongly as the building industry. The assurance of speed, efficiency, and guaranteed results lures the students to consume such products almost too comfortably.

Today, the rise of machine learning and big data draws attention to the critical usership of intelligent software products and data in a new light. Debates in the field of education and social sciences emphasize the importance of critical digital literacy[38] and digital empowerment.[39] Also, a growing body of scholarship and practices challenge the idea that science and/or technology is objective and neutral by demonstrating how scientific thought is situated in particular cultural, historical, economic, and social systems[40] and reveal preexisting negative (racial, sexist, gender) biases in digital culture, data science and algorithms.[41] Feminist principles are being incorporated in the technology and design-oriented fields such as Science and Technology Studies, Human–Computer Interaction, Digital Humanities, and Geography/GIS to draw attention to questions of epistemology.[42] Data journalism provides inspiring investigative uses of data either as the source or as a tool with which complex stories are generated and told through engaging infographics—or both.[43] Data Feminism (2020) suggests six core principles to be applied to data visualization and visual representations.[44] Some of the conceptualization of criticality in architecture with a focus on the digital production include critical design (Dunne 2005), critical making (Ratto 2011) and critical imagination (Cardoso Llach and Ozkar 2019).

Considering the global socio-environmental urgencies presented in the beginning, it is increasingly important to question how an architectural project benefits from and

---

his architectural practice frequently delayed by ever-changing drawings and fluctuating design comments (Weckherlin, 2007). He initiated his ideas as pedagogical projects in his teaching practice. In "Rapid Design" course, he created a standard card-index box of 'typical' design solutions to compile an 'open-source' catalogue for the students to copy and assemble in their projects. In the 'Active Building Atelier', designed projects were actually built to demonstrate the results of a design logic that is efficient in procedure and based on rationality (Weckherlin, 2007).

[38] Pangrazio, Luci. (2014). Reconceptualising critical digital literacy. Discourse: Studies in the Cultural Politics of Education. 37. 1–12. 10.1080/01596306.2014.942836.

[39] Tissenbaum M., Sheldon J., Seop L., Lee C. H., Lao N., Critical Computational Empowerment: Engaging Youth as Shapers of the Digital Future 2017 IEEE Global Engineering Education Conference (EDUCON).

[40] J. Wajcman. Feminist theories of technology. Cambridge Journal of Economics, 34(1):143–152, 2010. https://doi.org/10.1093/cje/ben057.

[41] *Algorithms of Oppression: How Search Engines Reinforce Racism, Safiya Umoja Noble, NYU Press, 2018.*

[42] *Feminist Data Visualization, Catherine D'Ignazio, Lauren F. Klein, Workshop on Visualization for the Digital Humanities (VIS4DH), Baltimore. IEEE. 2016.*

[43] "What is data journalism?" https://datajournalism.com/read/handbook/one/introduction/what-is-data-journalism.

[44] Six principles of Data Feminism: Rethink Binaries, Embrace Pluralism, Examine Power and Aspire to Empowerment, Consider Context, Legitimize Embodiment and Affect, Make Labor Visible.

contributes to the above-discussed critical digital social and creative practices and new epistemological frameworks that enable them.

Another emerging quality for an architectural project is the degree of the critical distance taken in its operations and outcomes. Was the selection of technology or design response debated with their consequences in mind? Critical distancing requires a redefinition of progress in spatial practices, which does not prioritize high-tech over the low-tech, fast over the slow, but the idiosyncrasies of each condition. In an online talk,[45] Jeremy Till quotes, "To a man with a hammer, everything looks like a nail" to problematize the professional attitudes in architecture. He argues how "the hammer is both a strength because it forges the zone of professional expertise, but it is also a weakness because the world is not just nails, hence the expansion of the range of available tools."

The question of distancing applies to the design response given to a spatial situation. Was a building really the best solution? There are spatial aspects of the society that do not necessarily materialize in the form of a new building, not even a prototype or a construction technique. Did the project distance itself enough to equally review possible non- or immaterial responses: such as the organization of information and communication for inhabiting the built environment or interacting socially in space? The habit of clinging to buildings as solutions to spatial problems calls for rethinking of spatial practices and dogmas in architectural pedagogy. Considering the role of construction to climate crisis, the need to rethink the capacity of architectural practice in the range of physical, corporeal, material, and digital spatial experiences is crucial.

How does the project deal with the "messy" or "wicked" nature of a spatial problem? Is it fixated on "solving" it, or highlighting problems to address? Awan, Schneider and Till (2011) discuss how limiting environmental understanding to the technical realm alone leads to a sense that environmental issues can be dealt with through technical fixes and gives a false sense of security because the environment is clearly tied into much wider networks. They argue for an engagement with these networks, which is not isolated to matters of energy reduction and efficiency, but has to be understood in relation to the social, global and virtual realms.

Lastly, how much does the project trigger or is triggered by spatial imagination? Does it construct new narratives and dream about new scenarios of spatial production, spatial intelligence and spatial agency dealing with the environmental catastrophes and social challenges? Is the digital infrastructure instrumentalized to connect spatial imagination with feedback systems, issues of labor, economy, material, social and environmental impact? How are the conceptualizing, storytelling, and visualizing tools employed to support new spatial imaginations (Fig. 10.3)?

These questions speculating on the new qualities for a good project compel the students as well as the curriculum designers and studio tutors to recharge design-learning agendas and environments with such provocations.

Overall, the essay attempted to question, speculate and provoke disruptive pedagogies for architecture, through the three historically charged, formative elements that

[45] https://www.youtube.com/watch?v=xCBYAezddg0, *retrieved April 10, 2021.*

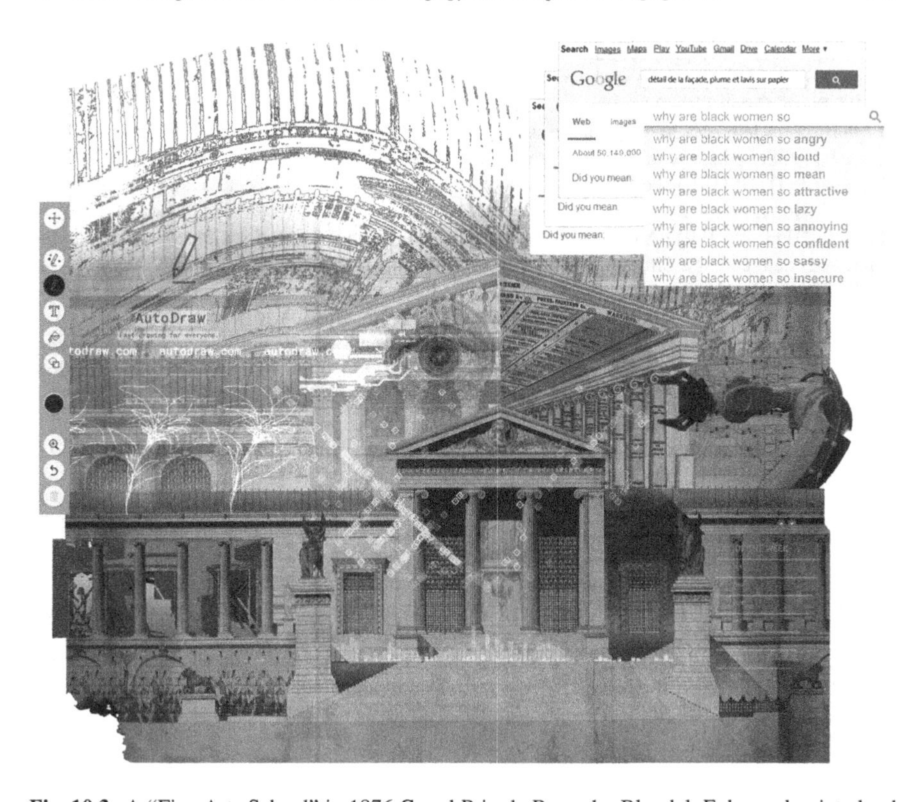

Fig. 10.3 A "Fine Arts School" in 1876 Grand Prix de Rome by Blondel, Enhanced point cloud visualization, Tekla Structures (2020), Algorithms of Oppression: How Search Engines Reinforce Racism (Umoja Noble, 2018), Emma Willard, Temple of Time (1846), Autodraw, Google's new AI experiment "autocorrects" doodles (2020), On Weathering: the life of buildings in time (Mostafavi, Leatherbarrow, 1992), Augmented Reality Contact Lens products. Collage executed by Anıl Aydınoğlu

continue to structure architectural learning experiences today. "What is it to learn for an unknown future?" (Barnett 2004) may be an old question. But it resonates much more vibrantly today in the face of unprecedented radical global changes that compel architectural education to restructure its pedagogies to embrace these disruptive forces for their emancipatory potentials.

References

Allen S (2012) The future that is now, architecture school: three centuries of educating architects in North America, edited by Joan Ockman, The MIT Press

Barnett R (2004) Learning for an unknown future. Higher education research and development, vol 23, No. 3

Carpo M (2011) The second digital turn, design beyond intelligence. The MIT Press

Carpo M (2017) The alphabet and the algorithm. The MIT Press
Dunne A (2005) Hertzian tales: electronic products, aesthetic experience, and critical design. The MIT Press, Cambridge, MA
D'Ignazio C, Klein LF (2020) Data feminism. The MIT Press, Cambridge, MA
Freire P (1971) Pedagogy of the oppressed. Seabury Press, New York
Gruenewald D (2003) The best of both worlds: a critical pedagogy of place. Educational researcher, vol 32, No. 4, pp 3–12
Haraway DJ (1985, 1991) A cyborg manifesto: science, technology, and socialist-feminism in the late twentieth century. Simians, cyborgs and women: the reinvention of nature, Routledge
Illich I (1995) Deschooling society. Marion Boyars Publishers Ltd., London
Llach DC (2015) Builders of the vision, software and the imagination of design. Routledge
Llach DC, Ozkar M (2019) Cultivating the critical imagination: post-disciplinary pedagogy in a computational design laboratory. Digit. Creat. 30(4):257–276. https://doi.org/10.1080/14626268.2019.1691604
May J (2017) Everything is already an image. Essay published in Log 40, Spring/Summer
Ratto M (2011) The information society 27:252–260; Taylor & Francis Group, LLC, ISSN: 0197-2243 print. https://doi.org/10.1080/01972243.2011.583819
Stevens G (2002) The favored circle, the social foundations of architectural distinction, The MIT Press
Smith E, Selfe CL (2006) Teaching and transformation: Donna Haraway's "A Manifesto for Cyborgs" and its influence in computer-supported composition classrooms. In: Weiss J, Nolan J, Hunsinger J, Trifonas P (eds.) The international handbook of virtual learning environments. Springer, Dordrecht. https://doi.org/10.1007/978-1-4020-3803-7_5
Weckherlin G (2007) Ernst neufert's architects' data: anxiety, creativity and authorial abdication. In: Anstey T, Grillner K, Hughes R (eds) Architecture and authorship. Black Dog Publishing, London, p 146

Part III
Cyber-Urban Integration, Tectonism, and Disruptions

Chapter 11
Cyber-Urban Integration, Tectonism, and Disruptions

Philippe Morel

11.1 Introduction

When we look at recent changes in architecture, let's say the last 50 years, between 1972 and 2022, the characteristic that strikes us most is the gap between the architecture commonly built in 1972—including its modes of practice that we would call today "business models"—and that built in the last decade. Remember that 1972 is an almost arbitrary date, although that same year *Learning from Las Vegas* was published by MIT Press. Although 50 years is a very short time in the history of architecture, construction and the city, and of course a short time in the history of technology as well, we feel as if we are just as separated from 1972 as we are from the age of the steam engine. Indeed, the debates of an era that now embodies the birth of postmodernism, or at least a form of culturalist postmodernism, seem to us today to be naïve, 'arty' and self-centered on the intellectual elite that produced them. These debates also seem reductionist, or at least very much out of step with the radicality of the transformations at work in the organization and planning of business at the global level: the computerization of trade and markets with the computerization of the NASDAQ in 1971, the explosion of container-based logistics, urban hyper-growth, the explosion of tourism, the end of the Bretton-Woods agreements from August 1971, etc. While from the end of the 1960s onwards certain architects (e.g. John Negroponte) and technologists tried, *by technological means*, to take better account of urban realities and the needs and wishes of the inhabitants, what will remain overall from this era will be social experiments in direct participation that were quickly rendered obsolete by the complexity and slowness of the decision-making processes, confronted by the speed and power of the market. Today, in a post-internet era that seems already

P. Morel (✉)
The Bartlett School of Architecture, University College London, London, UK
e-mail: p.morel@ucl.ac.uk

P. Morel and H. Bier (eds.), *Disruptive Technologies: The Convergence of New Paradigms in Architecture*, Springer Series in Adaptive Environments,
https://doi.org/10.1007/978-3-031-14160-7_11

formidably accomplished, even though it is only an embryonic state of a new civilization, there are new calls for an architecture that is fundamentally connected to reality—but to *the whole of reality*, not to its formal, stylistic, aesthetic, social or economic aspects taken individually. Some of these calls come from the authors of the chapters in this section *Cyber-Urban Integration, Tectonism, and Disruptions.* Vishu Bhooshan, Henry David Louth, and Shajay Bhooshan advocate the need for a new use of advanced technologies and a new form of provision of both the tools and the results of their use by architects; Philippe Morel reminds us of the need for new forms of practice and, to this end, for a better knowledge of the mechanisms of innovation. As for Patrik Schumacher, he is undoubtedly the practitioner and theorist whose work has the widest visibility, audience, and impact. While 'parametricism' has been perceived as a new attempt to restore a style, which its author has defended, arguing that only a style has the power to transmit a new set of values, no attentive reader can deny that the theoretical richness of this concept goes beyond this issue. Hence, the first of the three chapters by Patrik Schumacher (Zaha Hadid Architects, Architectural Association), entitled *Cyber-Urban Integration*, represents a further development of Schumacher's thinking. It speculates on the current integration of the digital and the physical within new *"cyber-urban"* environments. According to the author, *"after 30 years of theoretical speculation and advances in gaming and entertainment, the internet is finally on the way to transforming into cyberspace. The magazine as a guiding analogy for the web is being overtaken by the analogy of the city. Architects take over from graphic designers. The premise for the plausibility of this takeover and expansion of architecture's competency is that all design, including architecture, is communicative framing. The thesis of this paper is that in this age of soaring web-based telecommunication, the space of social communication must be designed simultaneously as a physical and virtual realm, as cyber-urban space, seamlessly integrating physically immediate and digitally mediated communicative interactions, constituting a new augmented mixed reality."* In his chapter, P. Schumacher elaborates on the nature of *"architecture's core competency"* through what he calls *"the four architectural projects"*. He shows how these projects are dependent on a new industrial system, a new *"pro-active Intelligent Environments"* and an agent-based parametric semiology that, according to him, should be expended to realize the full potential of finally mature cyberspace within the discipline of architecture and beyond. Such a cyberspace representing, according to Michael Benedikt whose 1991 book *Cyberspace: First Steps* is discussed by P. Schumacher, *"a new stage, a new and irresistible development in the elaboration of human culture"*.[1] The second chapter—*Democratising Tectonism: A high-performance technological basis for engaging and responsible design, online and on-land*—by Vishu Bhooshan, Henry David Louth, and Shajay Bhooshan (Zaha Hadid Architects, Architectural Association), deals with the possibility of such a democratization through the concept of *"Spatial Technology Stack (STS)"*, that unifies Architectural Geometry and game-tech. According to the authors, such an STS could *"robustly support the synthesis of high-performance shapes including structurally optimized geometry and its processing for robotic and digital fabrication (RDF), and the creation of environments that deliver novel, engaging and productive spatial user experiences both in the physical and virtual instantiations of architecture"*. Contrary to *"misaligned building information modelling technologies"*, the STS could finally provide *"an alternative high-performance technological basis for engaging and responsible design, both online and on-land*, within the context of a new *"cultural production view of architecture, spatial user-experience (UX) design and end-user ergonomics"*. The third and last chapter of the section, by Philippe Morel (Associate Professor at UCL Bartlett & ENSA Paris-Malaquais, initiator and founding CEO of XtreeE), entitled *Why Disruptive Business Models are Inseparable from Disruptive Technologies*, goes back to the importance of novel business models in today's technological explosion. It addresses the relationship between business models and disruptive technologies as a counterpoint to the general theme of this volume "*Disruptive Technologies: The Convergence of New Paradigms in Architecture"*. While discourse on disruptive technologies commonly insists on the technologies themselves, most often from the point of view of their technical operativity or from an epistemological perspective, a closer look at the reality of techno-capitalist societies reveals the crucial importance of how technologies are inserted into the global economic market. This insertion obviously impacts the technological appropriation, but maybe more importantly the technological evolution itself, including in architecture perceived here in a broad sense, from the conception to the maintenance of projects after delivery. By looking at a few arguments about the nature of disruptive technology and innovation, including from the inventor of that very notion of disruptive innovation, this final chapter demonstrates how different our present architectural time is from everything that preceded it. Indeed, while business models in architecture have rarely ever changed until the beginning of the XXI century, new models might become one of the most important parameters of change in the post-internet era, beyond mere technological change which is far too often the unique concern of architects.

Notes

1. Michael Benedikt (Ed.), Cyberspace: First Steps, MIT Press, Cambridge MA 1991.

Chapter 12
Cyber-Urban Integration

Patrik Schumacher

12.1 Architecture's Core Competency: The Four Architectural Projects

The life process of society is a communication process that is ordered via a rich typology of communicative situations. It is the designed environment, both physical and digital, that spatially distributes, frames, stabilises and coordinates these distinct situations within an evolving order that allows us to self-sort as participants of various specific social interactions. Designing is communicative framing. This insight must now be made the explicit premise and agenda for a systematic design research project that bridges architecture and interaction design in 3D virtual worlds that must at the same time connect up with our lived physical space.

The design of virtual communication spaces lies fully within the architect's core competency. Any design project in this space involves the three parts of all architectural projects distinguished in 'The Autopoiesis of Architecture'[1]: the organisational project, the phenomenological project and the semiological project. The semiological project is crucial: While all urban spaces are never only mere physical containers that carry and channel bodies but are always also information-rich navigation and semantically tagged interaction spaces, this information-rich, semantic charge and communicative capacity is, in the case of cyberspace, distilled as the very essence of all design efforts. Here there can no longer arise the confusion of the designer's task of ordering and framing social interaction with the provision of a physically specified shelter and its technical construction.

To design architectural projects, real or virtual, implies the development of a grammar empowered spatio-visual language, with a much enhanced communicative capacity, to create navigable and legible information-rich environments for

P. Schumacher (✉)
London, UK
e-mail: Patrik.Schumacher@zaha-hadid.com

P. Morel and H. Bier (eds.), *Disruptive Technologies: The Convergence of New Paradigms in Architecture*, Springer Series in Adaptive Environments,
https://doi.org/10.1007/978-3-031-14160-7_12

densely layered societal interaction types, each with their differentiated purposes and selectively gathered audiences.

The new task of cyber-urban integration brings forward a fourth project, the dramaturgical project, i.e. the architectural equivalence of what in web-design is called interaction design, as a key aspect of user interface design (UI) or user experience design (UX). The dramaturgical project also exists for building design, to the extent that users can interact with buildings, i.e. open doors, opening or closing curtains etc. This dramaturgical project will become much more prominent now, both within the virtual and the physical domains of interaction. Within the physical domain the author has worked on the possibilities of the dramaturgical approach to architecture via AI empowered kinetic architectural elements and systems. This agenda was pursued within the AADRL research project of 'responsive environments', and via a more recent update under the heading of 'spontaneous creative environments'. Currently this agenda is pushed further via the design research project of a cyber-urban incubator. Naturally, the implied continuous adaptive self-re-organisation of the framing environment can be accomplished much more effortlessly within the virtual domain.

12.2 A New Life in a New Industrial System

The built environment must progress in step with the progress of society. It is therefore the task of the avant-garde segment of the academic discipline and profession of architecture to theorise and explore how best to guide the development of the built environment in ways that are congenial to the opportunities and challenges of the technological and societal development at the frontier of progress. This requires that architectural theorists connect up with an updated theory of society and its probable trajectories of progress.

The new computationally empowered economy implies a shift from routine work to intensely collaborative work patterns. Nearly all work becomes creative work like R&D, marketing, and finance, together feeding a world of 3D printing, robotic fabrication and software as a service. The new reprogrammable robotic production technologies can absorb an unlimited number of innovations. There is no technical or cost limitations in uploading new improved apps to millions of users every day, or to feed 3D printers with new improved instructions. Also, robotic assembly lines no longer lock workers into routine work. All workers are set free to innovate. This should eventually allow everybody to become a self-directed creative innovator. This increased innovation absorption capacity of current production technologies implies a momentous intensification of communication and collaboration, since innovations require the re-integration of all the specialised aspects of a product or service. This means that most work will not only become creative, but intensely communicative, in science, R&D, design, marketing, media, finance, education, etc. This intensification of creative collaboration and communication implies a new level of urban concentration as well as a new level of cyber-spatial agglomeration.

Co-working and incubator spaces will make up an increasing part of the urban fabric, interlaced with spaces for more freewheeling networking and socialising. The idea of a Cyber-Urban Incubator is proposing to double up these urban realms with corresponding virtual realms, not as digital twin replicas but as congenial extensions with their own laws of navigation, encounter and interaction but integrated via spatial interfaces and via a unified spatio-visual language as a broadly shared orienting system. Physical spaces will afford windows into virtual spaces where the semiology is the same and the logic of gathering and communicating is similar enough to allow for the transfer of competencies from the real to the virtual realm. The easy implementation of data-rich informational overlays in the virtual city extensions might inspire ways to deliver such augmentations also within the physical city experience. In general it is likely that the age of cyber-urban integration will eventually also impact back not only onto the utilisation of urban spaces but on the spatial organisation of the urban itself. While this can be expected in the abstract, the concrete forms this might take are yet to be discovered.

12.3 Pro-active Intelligent Environments

The unprecedented level of dynamism in social interaction processes in contemporary creative industry work environments calls for adaptive, responsive and indeed creative built environments. The discourse of so called ‘intelligent buildings’ has to be radicalised and related to the core competency of architectural design, namely the ordering of social interactions. If these patterns of interaction become increasingly variable, this implies the demand for an unprecedented level of real time spatial flexibility. This demand can only be met by perceptive, responsive environments. However, the next step here is truly intelligent, creative environments that operate in a pro-active, self-directed fashion rather than merely responding in routine ways or waiting for instructions. The architectural elements or ‘agents’ that are meant to facilitate increasingly complex and dynamic patterns of human collaborative interaction must become congenial participants in the collective life process. Just as a contemporary tech firms consists of self-directed collaborators that develop their own initiatives rather than employees waiting for instructions, so will a future work environment consist of robotic agents that do not wait to be remote controlled but are self-acting and learning to maximise their usefulness and actual utilisation.

The scene is set, within contemporary advanced work environments, for the architectural instrumentalisation of the artistic experiments with interactive art installations, powered by the new easy availability of sensor and actuator technologies. Doors, windows, blinds, partitions, screens, tables, desks, chairs, coffee machines, water coolers, lighting devices etc. will all become self-directed agents, with a life-long machine learning curve, steered by the prerogative of maximising their inbuilt utility functions that guide them to increase their utilisation and usefulness in the social communication process. In both the real and the virtual spaces ai-empowered architectural agents and avatars will, more and more, become our valued collaborative companions. To the extent that such entities accumulate experiences and evolve unique skills and knowledge, due to their individual histories, they become irreplaceable and thus precious, not unlike human persons. The life-process of the future will thus become a man–machine ecology, with many productive human personalities and many more productive machine or system personalities.

This is the concept of spontaneously intelligent environments. In our post-Covid-19 world, these work environments will have to be seamlessly connected up with the virtual communication spaces for those who will participate remotely rather than via physical co-presence. This project of pro-active intelligent environments is naturally congenial with the project of cyberspace, and indeed the implementation of the idea of continuous adaptive self-transformation can now be spearheaded within virtual spaces due to the relative ease of its realisation compared with physical versions.

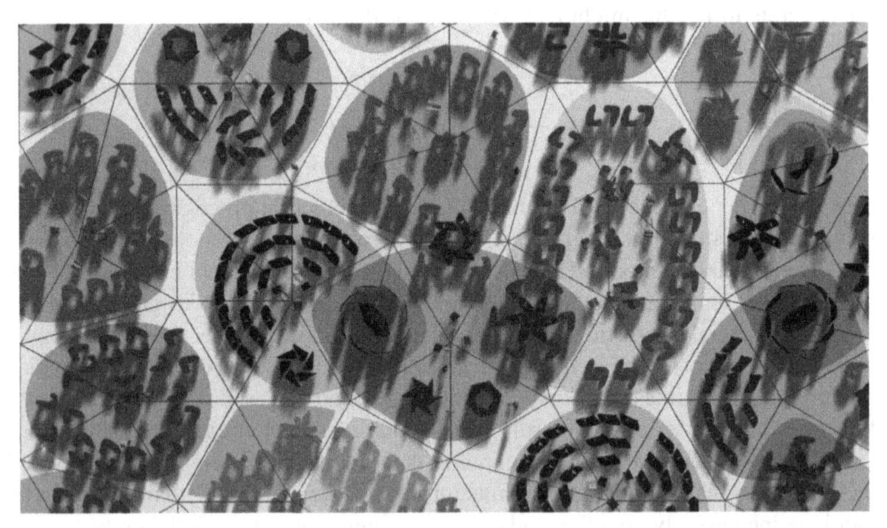

Robotic AI empowered self-directed work environment, AADRL, London 2020, Team: Man Mei Lam, Huiyuan Li, Ruixue Wang, Xuan Zhou, Tutors: Patrik Schumacher, Pierandrea Angius

12.4 Expanding Agent-Based Parametric Semiology

While every architect has an intuitive grasp of the normative interaction protocols that attach to the various designated areas that the design brief indicates and usually knows enough about the expected and desired user occupancy patterns, such intuitions cannot give a secure guidance on the relative social performance of alternative designs for large, complex environments. Intuition must here be substituted by simulations that can process thousands of agents interacting across an environment of hundreds of spaces. When quantitative comparisons and optimisation are aimed at, then intuition fails already in much smaller, simpler settings.

The simulation methodology developed under the research agenda 'Agent-based Parametric Semiology' is conceived as a generalisation and corresponding upgrade of the kind of crowd simulations currently offered by traffic and engineering consultants concerned with evacuation or smooth circulation. These crowd modellers treat users as physical bodies and simulate crowds like a physical fluid. In contrast, architectural design considerations are concerned with socialised actors who orient and interact within a semantically differentiated environment. The simulations that must be developed to get a handle on desired social interaction scenarios will have to be different and rather more elaborate. They contain circulation models as a relative trivial component. There are a number of crucial advancements that distinguish our architectural crowd models from the prior engineering models:

The first and most obvious advance is the expansion of the menu of action types. The second major advancement is that the agent population is socially differentiated rather than homogenous. For instance within the domain of corporate office life, agent differentiation might track rank, functional role, and team affiliation. An intricate network structure might be read off the client's intra-net to inform the agent population within the simulation. The third significant difference and upgrade is the dependency of the agents' behaviours on the functional designations of the spaces. The environment is zoned via designated and semantically encoded areas. Agents change their interaction propensity accordingly. This ordering increases the probability of highly specific interactions. Where the number of designations and protocols to be distinguished is very large, it is opportune to use the combinatorial power of grammar to articulate this manifold. This thus implies a fourth enhancement, namely the elaboration of an agent system with language competent agents.

The fifth aspect that distinguishes these architectural-semiological models from the circulatory crowd models is the following: Congenial with contemporary cultural conditions, the underlying presumption of these models is that agents are largely self-directed, rather than running on pre-scheduled tracks, and do self-select their interactions and the social events they participate in. These selections are guided by multi-dimensional, dynamic utility functions that can utilise contingent opportunities that the agents encounter within the environment they browse through. These utility functions are implemented in the decision processes that control the agents' actions on the basis of internal states due to prior actions and the environmental offerings perceived.[2]

This research project of 'agent-based parametric semiology'[3] started within AADRL, then migrated into the next development phase with a PhD group at the University of Applied Arts in Vienna. The next push was made within Zaha Hadid Architects. Our ZHA research team 'ZH Social' is currently moving from the research and experimentation phase to the implementation and testing phase within ZHA corporate headquarters projects.

The research team is building up increasingly large, differentiated and sophisticated agent populations using Unity game development software as base system, augmented with original coding. The development work concerning the agent populations benefits from a technology transfer from the game development industry, both with respect to basics like action animation and simple tasks like pathfinding, obstacle avoidance etc., and with respect to the more complex decision making processes modelled in 'game AI'. Sophisticated games populate their scenes with increasingly versatile, intelligent and spontaneous life-like agents.

The latest game AI methodology that is becoming more widespread in the game development industry is a methodology employing utility functions, so called 'Utility AI'. Instead of switching between a set of finite states based on conditioning via triggers, or moving through a whole decision tree until trigger conditions are met, in Utility AI agents constantly assess the actions available to them in their current environment, and assign a utility score to each of those actions on a continuous scale. The utility system then selects the behaviour option that scores highest amongst the currently available options, based on the circumstances. Circumstances are both external and internal states. The latter being dependent on what went on in the game or simulation so far, i.e. the current utility and thus the urgency of a desire. The utility of the related action recedes or drops after the action was successfully completed and the desire was satisfied. The basic laws of subjective economics like the law of diminishing marginal utility can be thus be implemented here. The normalised utility functions bring the most diverse and otherwise incommensurable measures into direct comparison. Each choice of action is relative rather than based on absolute conditionals. These are temporary prioritising decisions, based on internal states like desires, their urgency, available energy levels, as well as on opportunities afforded by environmental offering in proximity to current location. The various designated zones pre-condition the available action menu. Utility AI can take any group of action options, destination objects, interaction chances and score these. This makes the methodology very versatile for decision-making.

This technology transfer from the gaming industry delivers thinking tools, formalisation strategies and coding techniques for the elaboration of sophisticated autonomous agents capable of navigation and interaction within semantically charged environments.

The agent-based life-process models bring the interaction processes that are shaped by the designed architectural frames, i.e. the envisaged meaning of these spaces, into the design model. This way we are achieving the empowering operationalisation of both the semiological and the dramaturgical project. This constitutes a significant upgrade to our discipline's capacity to maintain a grip on social functionality in the face of an increasing complexity and dynamism within the built environment.

This methodology, and indeed the developed tool sets, are readily transferable and adaptable to the new task of designing virtual twins and virtual expansions for the new era of cyberspace and cyber-urban integration.

Indeed the same type of agent models ZH Social had been developing for the comparative testing and upgrading of an architectural project's social functionality are now being adapted to simulate and comparatively test the efficacy of designs for virtual interaction spaces, i.e. for the simulation and appraisal of interaction processes that are possible and can be expected in the virtual environments we are designing, as well as in the mixed reality spaces we are envisaging. Our development work with respect to life-process simulation is also contributing to the task of populating virtual environments with autonomous agents or 'virtual humans'. This desire and need originated in the gaming industry—the source domain of our technology transfer—but could also make sense in performance oriented environments, as proactive animators, or as a background population illustrating the social situation and social protocols that are meant to be facilitated in the respective space.

Zaha Hadid Architects (ZH Social), Agent-based life-process simulation for Sberbank Technology Centre, Moscow 2020

12.5 The Delayed Advent of Cyberspace

All the design disciplines, from urban design and architecture to fashion and graphic design, together do or should form a unified discourse and practice with a unity of purpose: the sensuous framing of communicative social interaction. This also includes all web design, all video-conferencing platforms, as well as all virtual collaboration platforms. Here too our colleagues' framing design work is always involved.

The internet started as a mainly academic network in the 1980s and took off more broadly in the early 1990s. Soon some of us architects imagined that the internet would develop into a virtual three-dimensional navigation and communication space, i.e. 'cyberspace'. The word "cyberspace" was coined by science fiction writer William Gibson, in his 1984 novel 'Neuromancer'. The design studio I was teaching at TU Berlin in 1995 was exploring this idea under the heading

'Virtual College': Online learning as collective experience facilitated within a virtual architecture. Informative inspiration was drawn from architect Michael Benedikt's seminal book, first published in 1991: 'Cyberspace: First Steps'.[4] Benedikt mused about "a new stage, a new and irresistible development in the elaboration of human culture"(p. 1) and did speculate conscientiously and resourcefully about "the nature of the artificial or illusory space(s) of computer-sustained virtual worlds." (p. 119).

However, the internet became a magazine-like medium instead, the preserve of graphic designers rather than architects. This will change. Cyberspace is now firmly on the agenda.

Due to the long drawn out Covid-19 lockdown experienced across the world in 2020/2021 all communication, work collaboration, and all social events were pushed online, into the realm of digitally mediated interaction. The adoption of video-conferencing tools shot up massively, and so did the investment into this domain. We are currently witnessing an explosion of start-up companies offering virtual event spaces. This new situation accelerated a process that had been going on for a while. But mass adoption brings a wholly new dynamic into this realm.

This re-emergence of the idea of cyberspace, this time with accelerating practical pressure and much more technological power than 25 years ago, was rather sudden. Michael Benedikt's book remains a valid resource of inspiration.

Benedikt asks (and gives answers to) the key questions that remain relevant: "How might it (cyberspace) look like, how might we get around in it, and, most importantly, what might we usefully do there?"(p. 19). The last of these most general questions should probably be answered like this: We would want to do there everything we are doing in urban and architectural spaces: browse, communicate, work, learn, create, both individually and collaboratively, play, socialise, entertain etc. etc. The lockdown has impaired all urban and architectural interaction spaces and thus calls for everything to go virtual. This is a radically new situation. In the intervening years virtual environments were a choice, not a necessity, and the choice in favour of VR was made primarily in the realm of entertainment, especially via video games. This market had grown sufficiently large to deliver development resources, ample user market feedback, and a whole competitive industry. The fruits of these investments can now be reaped via technology transfer into societal domains where serious productive work is to be facilitated for adult users who have no time to waste. The forced push due to Covid-19 has led to the discovery that remote, mediated collaboration can be effective. This lesson cannot be unlearned and a new working lifestyle will emerge. The thesis of this paper is that this new life will be based on cyber-urban integration.

Benedikt asks further: "Which axioms and laws of nature ought to be retained in cyberspace, on the grounds that humans have successfully evolved on a planet where these are fixed and conditioning all phenomena (including human intelligence), and which axioms and laws can be adjusted or jettisoned for the sake of empowerment." (p. 119).

This is an important question, and there are many possible answers. In any event, cyberspace will have a "geography, a physics, a nature, and a rule of human law." (p. 123).

Benedikt shares some useful considerations and proposes some heuristics he discovered in the speculative cyber-space design explorations he conducted with his students. He rightly suggests that when cyberspace takes off "there will likely be myriad places in, and many regions of cyberspace—each with its own character, rules and function." He also anticipates that there will be a number of different competing kinds of cyberspaces, "each with its own culture, appearance, lore and law."(p. 122).

Benedikt introduces some useful basic distinction, like the distinction 'navigation versus destination', and the distinction 'extrinsic versus intrinsic' dimensions. These are dimensions of information encoding or visualisation, whereby the extrinsic dimensions are the two or three spatial dimensions that define an object's location or position in space (with time being a fourth extrinsic dimension) while an unbounded number of morphological properties or features are brought under the notion of intrinsic dimensions that might be used to distinguish and characterise an object or place in cyberspace. The important insight is put forward here that, with respect to the function of information conveyance, extrinsic and intrinsic encodings are in principle functionally equivalent, so that it is the cyberspace-designer's choice which aspect or information to encode via extrinsic variables, i.e. (absolute or relative) location/position, and which via intrinsic variables, i.e. shape, colour, materiality etc. The presumption here is—just as in the case of an urban order—that spatial positions are not randomly allocated but mean something and thus convey some (at least probabilistic) information about the actors and activities to be expected at the respective position.

While Benedikt does not reference architectural semiology, probably because he conceives cyberspace more in terms of data-visualisation than in terms of architecture and spaces of interaction, it became clear to me when I read 'Cyberspace' in the early 1990s that cyberspace design is essentially an effort in architectural semiology. I soon left my engagement with cyberspace behind (because the web became instead the domain of graphic designers) but my keen interest in the semiological project as a central aspect of the architect's core competency remained. With this came the theme of 'information density' which was also one of Benedikt's central themes for cyberspace design. The other theme that I brought back into architecture and urban design is the theme of orientation and navigation. Now my renewed engagement with the problem and task of cyberspace design brings me back full circle, well prepared for the challenge.

The distinction of navigation and destination is not a strict one. Most urban and architectural spaces are both navigation and destination spaces. The differentiation of pure navigation spaces like corridors, highways and subways are a modern phenomenon, but even these spaces are never wholly devoid of information and communication potentials but can offer more than the mere transition from A to B. The city can and should be browsed, and this browsing should also be a keen mode of engagement with cyberspace. We cannot assume that users know about all the offerings in advance but rather they must be enabled to browse, scan and discover what is there, not utterly randomly but in a structured browsing or search, where serendipitous discovery is enabled without a loss of overall orientation. Virtual environment researchers R. Darken & B. Peterson make this point too: "Navigation is rarely, if ever, the primary task. It just tends to get in the way of what you really want to do. Our goal is to make the execution of navigation tasks as transparent and trivial as possible, but not to preclude the elements of exploration and discovery. Disoriented people are anxious, uncomfortable, and generally unhappy. If these conditions can be avoided, exploration and discovery can take place."[5]

The surplus navigation can bring as an alternative to just jumping to pre-selected destinations, and has its equivalent in the slackness of lingering time around scheduled events. These informal pre-gatherings and the post-event lingering are very important for networking and informal 'browsing' information exchange. These networking processes make productive use of the non-random, select group brought together by the respective scheduled event, e.g. by a lecture, conference or exhibition opening etc. The utilisation of such an opportunity for explorative encounters and information exchange requires structured spaces of extended co-presence that are not available via conferencing tools like zoom, or in virtual exhibitions, both still based on the magazine page analogy rather than the city building analogy.

To return to Benedikt's question which axioms and laws of nature ought to be retained in cyberspace: The same question is posed with respect to the familiar organisation and articulation of the city, its spaces and of the buildings within it. How much of this must be retained in order to effectively exploit the city analogy, thus utilizing our familiarity with cities and our collectively shared competency as city dwellers and users of the panoply of building types and types of spaces that order our interactions in real space? The 'laws of the city' are much richer than the laws of nature. They are not universal a priori constraints but have co-evolved together with the societies they sustain, and must be understood historically, as embodying a historically transient pragmatic rationality.

While Benedikt presciently predicted the currently emerging virtual worlds and meta-verses when he talked about cyberspace as "a new universe, a parallel universe created and sustained by the world's computers and communication lines" (p. 1), my emphasis is on the integration and indeed fusion of real and virtual spaces.

When tasked with the simultaneous design of both the real and virtual spaces for a client the question also becomes: To which degree will the virtual extensions of the architecture retain the look, feel and logic of the real spaces? Probably to a very large extent, especially if we allow the new design features motivated by the modus operandi of the virtual expansion to feed back into the design of the spaces of real co-presence. Even if the dramaturgy is different, the semiological system of signification should be largely the same and cross the divide between real and virtual spaces.

This feedback or influence of the virtual design into the physical design should include attempts to physically implement the kind of pro-active adaptive mobile architectural agents I presume will be pioneered more pervasively in the virtual domain. The virtual domains will also effortlessly advance additional (real time) graphic information overlays. These too should, as much as possible, be implemented in the design of the physical interaction domains, via Google/Facebook glasses, via projections, or if no real time variability but only static information is applied, via further permanent morphological or material encoding. The presumption and promoted heuristics here is the massive increase in information density, both in the virtual and in the physical spaces, far beyond what we are used to encountering in architecture and urban design up to now. The hypothesis and hope in this respect is that the advent of cyberspace will lead to a new flourishing of architectural semiology. This is plausible or can be expected to the extent to which cyberspace will, from the perspective of its users, surpass any known city in terms of its variety and density of differentiated, effective interaction offerings. For this density to remain navigable, semiological articulation will become necessary. Large proprietary complexes or districts will probably be semiologically integrated by their dedicated or coordinated designers while larger agglomerations will engender a spontaneous semiosis that then feeds on itself in its further expansion and densification. In any event, architectural semiology, as the (still largely unacknowledged) essence of the cyberspace design task, has a better chance to succeed in cyberspace than in real space, not least due to the fierce global borderless competition in cyberspace, and due to the attendant more rapid historical turnover and remodelling of spaces. The increased communicative capacity that will then increasingly be expected by the users of cyberspace will lead them to expect or demand a similar information richness and communicative capacity from the physical urban and architectural spaces they are willing to patronise. The users' expectations and the competency in information absorption they acquired in cyberspace will fuel and finally force the semiological upgrading of the physical environment too.

This physical environment will not only acquire a new semiological density and coherence but will be transformed in many further respects as it gets enveloped by and infused with virtuality. Most walls and architectural and urban surfaces will become windows into virtual extensions connecting real to virtual spaces. Room-sized, full or partially enveloping panoramic screens or projections are very effective mechanisms of collective immersion into virtual spaces. Whole groups of physically co-located participants can thereby be tele-transported into a virtual environment, and thereby interact with several other groups. Another potent form of tele-presencing is holograms. The required equipment could be built into strategic locations like at the lectern in a lecture theatre. Both technologies are being advanced rapidly to ever greater effect and are ever more affordable. A further compelling technology for tele-presencing is Microsoft's VROOM—Virtual Robot Overlay for Online Meetings. Here telepresence robots allow remote users to freely explore a space they are not in, and provide a physical embodiment in that space. Here a robot acts on behalf of a remote participant in a real space as would an avatar in a virtual space. That robot is either equipped with a screen at head height to deliver a video presence of the remote participant, or becomes the site of an AR overlay for co-present participants wearing AR glasses. Holograms might also be spawned. These examples in hardware evolution imply that we must not imagine that cyberspace will be experienced only at home from a laptop, phone or headset, but within new types of technology empowered immersive spaces.

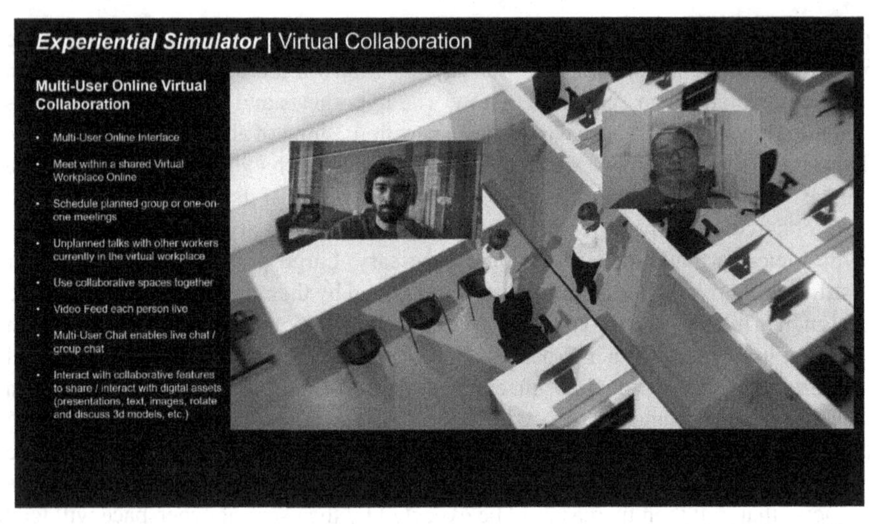

Zaha Hadid Architects (ZH Social), Digital Twin Cyberspace of ZHA Beijing Office, London/Beijing 2021

12.6 The Design of Cyber-Urban Incubators

The design of Cyber-Urban Incubators is underway. The idea of a cyber-urban incubator for the simultaneously virtual and real co-location of knowledge economy entrepreneurs serves as a test bed for the general agenda of cyber-urban integration. The cyber-urban incubator is ultimately meant to be the designed unity of real and virtual spaces for a corporate headquarter or campus, or for the larger branded knowledge industry incubation cluster. Zaha Hadid Architects have a number of relevant real projects under way and we are currently approaching our respective clients with the proposal to develop a virtual collaboration space together with our design of their physical premises, indeed mirroring our campus design proposals and transposing them into digital twin virtual interaction spaces. This allows us to utilise our 3D models as a convenient base for an online VR implementation ahead of real construction. This allows us to launch, market and test the design much sooner. This would also deliver a useful occasion to gather user and utilisation data, assuming that the virtual occupancy and utilisation patterns allow us to draw inferences about possible design improvements of the projected physical premises. At a later stage we imagine that the virtual spaces will take on their own developmental dynamic, however, without losing all continuities that are implied by the fact that both real and virtual environments are inhabited simultaneously by the same organisation, and moreover are tied together through many mixed reality scenarios where real/present and virtual/remote spaces and participants are operating jointly.

Candidate project for the implementation of a Cyber-urban Incubator: Zaha Hadid Architects, Tencent Technology Campus in Xian, London 2020/2021

Notes

1. Patrik Schumacher, The Autopoiesis of Architecture Vol.2 – A New Agenda for Architecture, John Wiley & Co., London 2012.
2. Patrik Schumacher, From Intuition to Simulation, in 'Positions: Unfolding Architectural Endeavors', Edition Angewandte, Birkhaeuser, Basel 2020.
3. Patrik Schumacher, Operationalising Architecture's Core Competency—Agent-based Parametric Semiology, in: DC I/O 2020, Design Computation Input/Output Conference: Algorithms, Cognitions, Cultures, London 2020.

   See also: Patrik Schumacher, From Intuition to Simulation, in 'Positions: Unfolding Architectural Endeavors', Edition Angewandte, Birkhaeuser, Basel 2020.

   See also: Patrik Schumacher, Advancing Social Functionality via Agent Based Parametric Semiology, in: AD Parametricism 2.0 – Rethinking Architecture's Agenda for the 21st Century, Guest-edited by Patrik Schumacher, AD Profile #240, March/April 2016.
4. Michael Benedikt (Ed.), Cyberspace: First Steps, MIT Press, Cambridge MA 1991.
5. Handbook of Virtual Environments: Design, Implementation, and Applications, Second Edition CRC Press, Taylor & Francis, London, p. 468).

Chapter 13
Democratising Tectonism: High Performance Geometry for Mass-Customisation of Virtual and Physical Spaces

Vishu Bhooshan, Henry David Louth, and Shajay Bhooshan

13.1 Democratising Impactful Digital Design and Construction

Architecture, like video games, movies, and music, is a technology-enabled cultural production. Architecture is not a product that is a direct outcome of technology. In other words, architects, engineers, and constructors use technological tools to realise ideas that are culturally and socially engaging. Games, the computer- generated movie industry and associated production pipelines have long understood this. Movie creators wield technological tools to tell socially and culturally engaging stories. Technical developers and the vast research industrial complex of the game and movie industry create the technological tools, not the movies. We ought to recognise this when applying software to architecture and construction.

Unlike the widely held belief in the Architecture Engineering and Construction (AEC) industry, problems in housing and other forms of socially driven and engaged urban development will not be solved by automation and vertically integrated project delivery alone. They could however be solved by democratising good design—creating the interactive design-assisting tools, incentivising user generated content, creating spatial pre-sets, adaptive components derived from high-performance (spatial & ecological) global best practice that is being demonstrated in the professionally generated content (PGC) developed by the bleeding edge architectural and engineering firms.

V. Bhooshan (✉) · H. D. Louth · S. Bhooshan
Zaha Hadid Architects Computation and Design Group (ZHA CODE), London, UK
e-mail: Vishu.Bhooshan@zaha-hadid.com

H. D. Louth
e-mail: Henry.Louth@zaha-hadid.com

S. Bhooshan
e-mail: Shajay.Bhooshan@zaha-hadid.com

P. Morel and H. Bier (eds.), *Disruptive Technologies: The Convergence of New Paradigms in Architecture*, Springer Series in Adaptive Environments,
https://doi.org/10.1007/978-3-031-14160-7_13

In such a context, *Architectural Geometry* (Pottmann 2010) (AG) is a highly relevant design technology paradigm. AG focusses on the synthesis of shapes that guarantee structural and fabrication optimality. It is also closely aligned with and complementary to the development of robotic and digital fabrication (RDF). Further, in combining historical geometry-based methods of structural analysis, modern mathematics as used in computer graphics (CG) and computational technologies, the field is opening several rich shape-possibilities that are also economically viable—a new *Tectonism* (Schumacher 2014a). Design that is so digitally empowered is already proving to be significantly more effective in terms of spatial expressivity and user-experience (Schumacher 2014b), ecologically (Rippmann et al. 2018), preservation of building trades (Fallacara 2009) etc. Thus, the recent and rising popularity of AG is not surprising considering it has brought the principal stakeholders in the architectural design process—architects, engineers and fabricators, and their respective toolchains much closer together (Louth et al. 2017; Bhooshan et al. 2018a).

Architectural Geometry, Tectonism and the Metaverse

Cyberspaces are virtual spatial environments in which human-to-human communication can happen, over computer networks. Current photo-real, high fidelity and massively multiplayer online (MMO) video-game creation technologies combined with high-speed network and cloud technologies enable such cyberspaces to be 3-dimensional (3D), interaction-rich and socially and sensorially engaging. Thus, cyberspaces are 3D spatial, digital assets augmented with communication capabilities. They are accessible from a variety of commercial devices including desktop browsers, mobile apps, smart TVs etc. Together, these features of cyberspaces make them an integral part of the spatial-web technologies underpinning the so-called *Metaverse*—a rapidly expanding online, socio-economic market, enabling novel cultural, social, and business opportunities.

However, at the heart of this exciting and vast architectural opportunity lies a technological divergence. Architects are not aware of game-tech used to create spatial content for the metaverse. On the other hand, game and so-called level developers are not aware of architectural design—tectonic, spatial, user experience logics and their implications on the look and feel of spaces. Thus, often the architecture depicted in the metaverse is bland or comical.

The explosive opportunities for both new business and rapid testing and refinement of core competencies of design provide the incentive to align architectural design technologies with those in game production. Unlike the Building Information Modelling (BIM) paradigm that is dominant and widespread in the AEC industry, Architectural Geometry is much better aligned with geometry processing, computer aided shape design, user-experience analytics, and other computer graphic technologies (Bhooshan 2016a, b, 2017). Thus, AG and game-tech are very compatible in terms of the underlying technology, in the designer-friendly, interactive design ethos and in supporting the dramaturgical focus of the creators.

Architectural Geometry, Games, and Governance Technology

Geometry-based abstraction of complex structural and manufacturing phenomena is an integral feature of AG. This means that many of its core technologies are easily transported to non-expert computer aided design (CAD) platforms such as web browsers, web-services, and game platforms. Gaming platforms are increasingly considered for 'gamified' solutions that require social engagement, multi-stakeholder participation, and negotiation of trade-offs (Bhooshan and Vazquez 2020).

Governance for the built environment requires such solutions. However, it is typically the anti-thesis of participatory solutions involving centrally instituted policies and regulations related to real estate taxes and policies, zoning laws, infrastructure development plans etc. By contrast, the so-called Governance Technologies (Govtech) are a technological layer for enabling effective, decentralised, participatory governance—allocation of resources, decision-making, and delivery of services to inhabitants. The stack of such technologies including block-chain, Internet of Things, decentralised finance, and their combinations with game-engine based, user-centric, interactive 3D platforms are gaining significant momentum.

Together, such cyber-physical platforms couple the social, exploratory and network-effect benefits of online 'metaverses' and the effective resource utilisation of digital twin technologies. On the one hand, they provide a minimal risk, online environment for experimentation, incorporation of participatory wisdom of non-experts, expert knowledge systems, and stakeholder freedoms. On the other hand, they provide expedient and resource efficient physical realisation and operation. Cyber-physical architecture and urbanism empowers human betterment via effective resource utilisation. They are imminent, exciting, and critical to the future of our societies and their physical receptacles.

A Technological Thesis Borne from Practice and Collaboration

Zaha Hadid Architects (ZHA) have collaboratively evolved expertise and proprietary technologies at the intersection of spatial user-experience, interaction, and design with computational technologies of algorithmic 3D geometry creation, game3D & MMO, and user-analytics. Furthermore, ZHA has early-adopter, pilot project experience in developing user-experience focussed spatial designs, and in the preparation of corresponding high quality 3D assets that are compatible with video game engines that power the high fidelity metaverse. ZHA also has long-standing expertise in developing 3D spatial and architectural assets by adapting so-called Digital Content Creation (DCC) toolsets that are commonly used in the computer graphics, animation, and video game industries. In fact, ZHA spent close to two decades collaborating and learning to extend the CG and game-development technologies stack for architectural production. Recent cyberspace-design collaborations with Player Unknown's Battlegrounds (PUBG), Kenny Schachter (NFTism) and Mytaverse

(Cyber-Liberland), validate these benefits of adapting game-tech (see Sect. 13.4). The following observations and a technological thesis follow from the experience and expertise accrued.

A so-called *Spatial Technology Stack* (STS), can robustly support the

- synthesis of high-performance shapes including structurally optimised geometry and its processing for robotic and digital fabrication (RDF) (Block et al. 2020; Block 2016);
- creation of environments that deliver novel, engaging and productive spatial user experiences (Schumacher 2014b)—both in the physical and virtual instantiations of architecture.

Stemming from the observations above is a technological thesis: *The spatial technology stack, compared to the current and dominant BIM-based architectural technologies, provides a powerful technological basis for engaging and responsible design, both online and on-land.*

We will argue, for the rest of the article, that STS is better aligned with

- The cultural production view of architecture and thus the core social and physical tasks of architecture.
- Spatial user-experience (UX) design and end-user ergonomics.
- Integrated design and construction and the ecological benefits thereof, including making them more widely available to the AEC industry.
- Game-tech powering the metaverse including so-called synthesis of performance optimised geometries, no-code or low-code platforms and application programming interface (API) requirements, collaborative technology development etc. This compatibility also enhances the potential to attract new talent into the AEC industry and to empower them to increase architectural experimentation via user-generated content (UGC).

13.2 Spatial Technology Stack

Spatial Technology Stack (STS) is the convergence of spatial design disciplines with computational technologies associated with architectural geometry (AG), computer graphics (CG) and gaming.

STS, which is congenial with Tectonism, incorporates and stylistically heightens the essential aspects of structure and fabrication in addition to increasingly encoding the social, ecological, and economic parameters into the shape modelling process. The recent advances and increasing popularity of AG have brought the principal stakeholders in the architectural design process—architects, engineers and fabricators, and their respective toolchains much closer together (Pottmann 2010; Panozzo et al. 2013; Prévost et al. 2013; Schwartzburg and Pauly 2013; Tamke 2015; Jiang et al. 2014; Michalatos and Payne 2016; Bhooshan et al. 2018b). This is true both in the early design phase and across the design to the physical production pipeline (see Sect. 13.3).

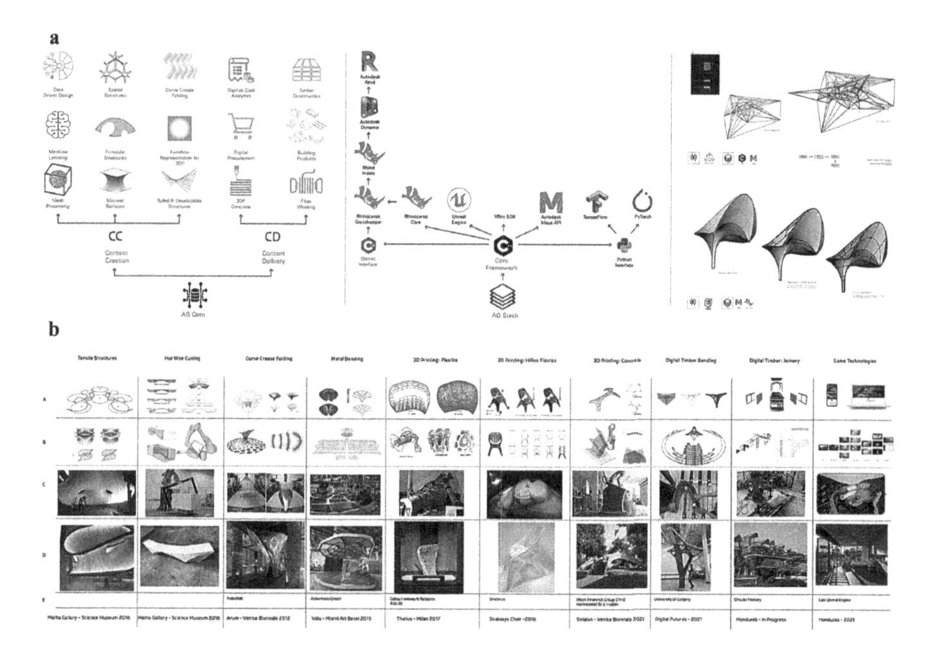

Fig. 13.1 **a** Development of technology stack to create toolsets, interoperability, and sand-box testing. **b** Digital and physical technology demonstrators

User Generated Content (UGC), typically used in the realm of journalism and social media, is disruptive as it empowers creative individual users with digital technologies (Lobato et al. 2012). In AEC, we believe STS in combination with curated professionally generated components will play a significant role in facilitating participatory practices and alleviate some of the critical constraints of only Professionally Generated Content (PGC) noted in Sect. 13.4. The Producers of PGC can thus turn "digitally empowered interactive audiences into value-generating co-producers" (Bruns 2007; Jenkins 2006). The recent developments in configurators, phygital spaces and metaverses further reinforce this trend (see Sect. 13.4) (Fig. 13.1).

13.3 Professionally Generated Content

The tech-stack and the interactive design environment (IDE) for Professionally Generated Content (PGC) inherits toolkits commonly found in the CG industry which enables the creation and manipulation of discrete representations of geometry—meshes, graphs, voxels—texture packing etc. Such discrete representations, though ubiquitous in the CG and animation industry, have hitherto not been as prevalent in architectural design. This is mostly due to the lack of appropriate creation and manipulation toolsets in popular Computer-Aided-Design (CAD) applications used by architects (Pottmann et al. 2006). Aided by recent developments in the application of discrete differential geometry to architectural design problems, the paradigm of AG favours discrete representation (Panozzo et al. 2013; Prévost et al. 2013; Schwartzburg and Pauly 2013; Tamke 2015; Jiang et al. 2014; Michalatos and Payne 2016; Bhooshan et al. 2018b). The Authors have invested more than 15 years in collaborating and learning to extend the CG and game development stack for IRL architecture. For further reading we point the reader to Louth et al. (2017; Bhooshan et al. 2015, 2018a, b, c, d, 2019; Bhooshan and Sayed 2011; Bhooshan 2016c; Reeves et al. 2016).

IDEs are valued as they facilitate the use of contemporary paradigms of edit and observe / interactive modelling (Prévost et al. 2013; Jiang et al. 2014; Rabinovich et al. 2018; Tang et al. 2014; Bhooshan et al. 2018c); exploration of static equilibrium shape design (Tang et al. 2014; Vouga et al. 2012; Block and Ochsendorf 2007; Akbarzadeh et al. 2015; Lee 2018); greater design control in the production and delivery stages (Louth et al. 2017; Bhooshan et al. 2018d) and provide a feedback loop between the various stages of the design workflow (Louth et al. 2017; Bhooshan et al. 2018a).

Benefits of Spatial Technology Stack

The following case studies illustrate the benefits of using STS in all phases of architectural projects—design, structural coordination, fabrication, and construction phases—as well as engaging the principal stakeholders of the projects.

Winton: The Mathematics Gallery at Science Museum, London

The design of the gallery, which welcomed more than a million visitors in the first six months and had increased dwell time in comparison to its predecessor, highlighted the benefits of considering rich spatial interaction-based user experience in the initial stages of design. Embedding analytics, associated UX metrics in an IDE, allowed for ease of iterating, and accommodating changes to the spatial layout if the objects, stories, or any other aspect of the curatorial vision were to change.

In addition, the IDE powered with integration of mathematical models, structural and fabrication constraints enabled:

- exploration of a wide variety of shapes in the constraint design space and negotiation of often disparate requirements—curatorial vision, ease of navigation, construction costs etc.
- exploration and iterative collaborative refinement of the fabric seams with the fabricator.
- effective workflow wherein the design could be updated and refined till the day of the production.

- customisation of each of the 14 benches to be unique, whilst still not compromising on the production time.

For more details regarding the project, we point the reader to Bhooshan (2016c) (Fig. 13.2).

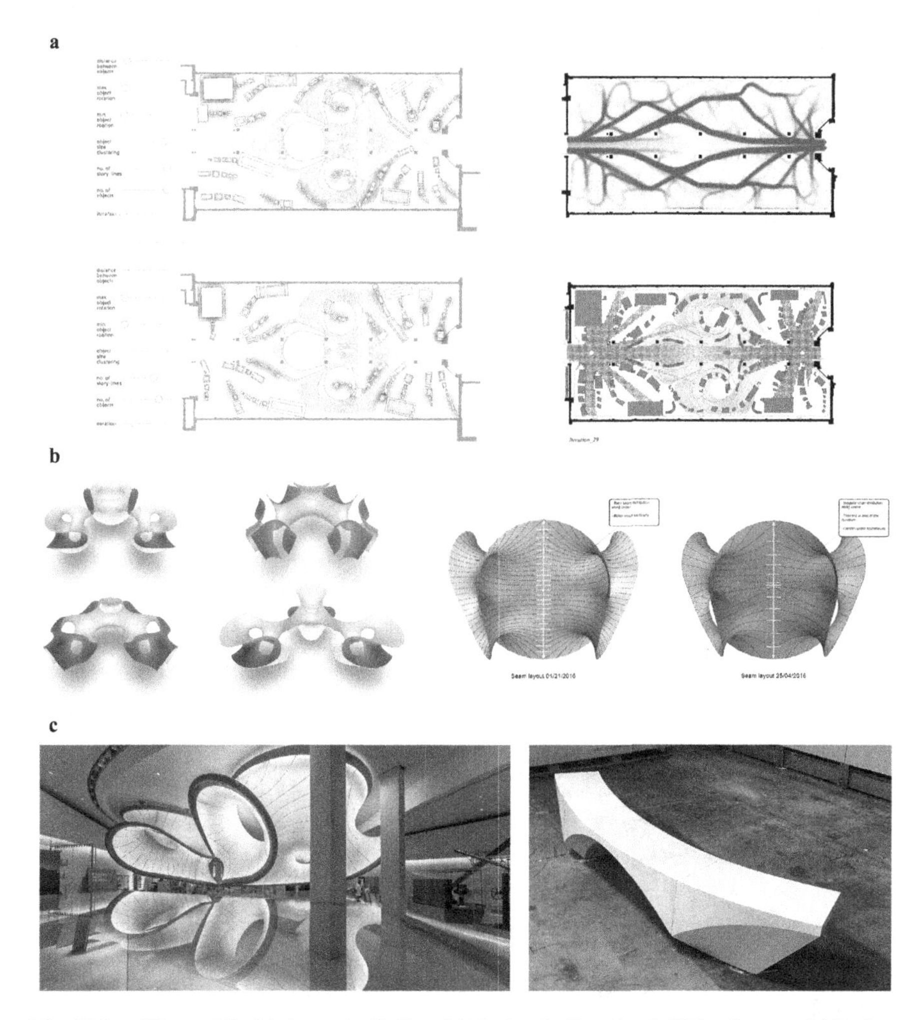

Fig. 13.2 **a** Winton: The Mathematics Gallery (a) Toolset for Parametric Object Layout, (b) Toolset for user parametrics and visual field analytics. **b** Winton: The Mathematics Gallery (a) Generative shape generation, (b) iterative seam pattern development of fabric pod. **c** Winton: The Mathematics Gallery (a) fabric pod, (b) cast, ultra-high-performance concrete benches

Striatus—3D Concrete Printed Masonry Bridge

Striatus is an arched, unreinforced masonry footbridge composed of 3D-printed concrete blocks assembled without mortar. The paradigm of strength through geometry coupled with precision placement of material only where needed using Robotic 3D printing, significantly reduced the environmental footprint of the bridge. Built without reinforcement and using dry assembly without binders, the bridge can be installed, dismantled, reassembled, and repurposed repeatedly; demonstrating how the three Rs of sustainability can be applied to concrete structures.

- Reduce: Lowering embodied emissions through structural geometry and additive manufacturing that minimises the consumption of resources and eliminates construction waste.

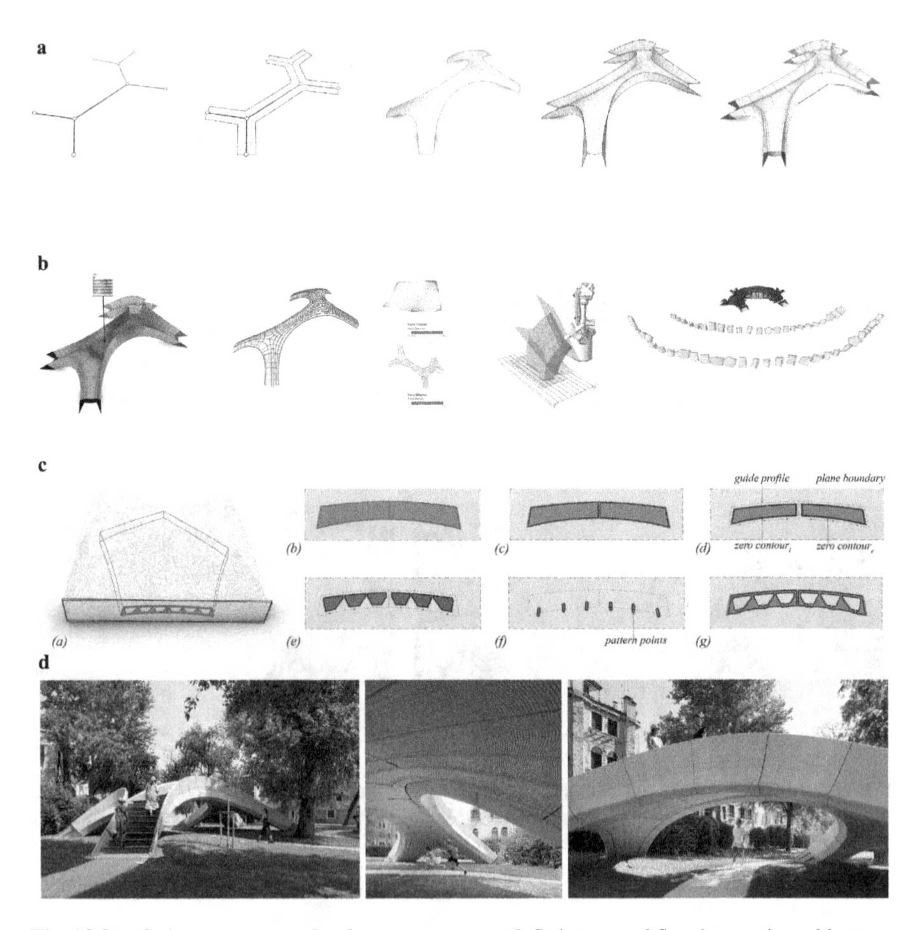

Fig. 13.3 **a** Striatus, geometry development sequence. **b** Striatus, workflow integration with structure, robotic fabrication and assembly. **c** Striatus, force aligned cross section and print path generation using sign distance fields. **d** Striatus, photographs of completed bridge

- Reuse: Improving circularity and longevity. Unlike conventional reinforced concrete structures, Striatus is designed to be dry assembled without any binder or glue, enabling the bridge to be dismantled and reused in other locations.
- Recycle: By ensuring varied materials are separated and separable, each component of Striatus can easily be recycled with minimal energy and cost.

The design-to-production (DTP) toolchain of the project developed using a mesh-based geometry-processing paradigm enabled for a collaborative and multi-authored design iteration and development. The use of JavaScript Object Notation (JSON) enabled lightweight and efficient transfer of 3D information with custom attributes between the various collaborators (Fig. 13.3). For more details regarding the project, we point the reader to Bhooshan et al. (2022a, 2022b).

Xi'an International Football Stadium

Set to be built in the Fengdong business district of Xi'an, one of the key design features of the stadium was the sheltering of the spectator seating with the lightweight large span roof. The shape design of the dual layer cable-net roof negotiates multiple objective constraints of spectator shading, natural light requirements for grass growth on the pitch, uniform force distribution, number, and length of cables. The integration of geometry-based form finding tools (Block and Ochsendorf 2007; Schek 1974) in the IDE during the initial stages of design enabled testing of multiple topologies and collaborative discussion and negotiation with the structural engineers for reduction of structural elements and depth especially at the oculus. The project highlights maturation of ST and its adoption in large scale creation and coordination of architectural packages—facades and envelopes, structure etc.—through efficient transfer of data streams using Geometry Method Statement (GMS) and associated parametric methods (Fig. 13.4).

Dnipro Metro Stations

The design of the station entrance shell canopies at Dnipro explored procedural generation of shapes from input graphs, which were adapted to the urban site conditions, entrance access, spatial, navigational, and other design constraints. The IDE helped design shapes which negotiated multi objectives of form-found surfaces with that of being developable for manufacture with flat sheet material. A streamlined parametric workflow between the various collaborators—architects, structural engineers, fabricators—enabled the delivery of all the station canopies—three stations, six canopies—concurrently and in a resource effective manner.

In summary, PGC powered by STS, is proving to be significantly more effective in terms of spatial expression and interaction rich user experience; ecologically—high performance with less material; efficient and lightweight exchange of data

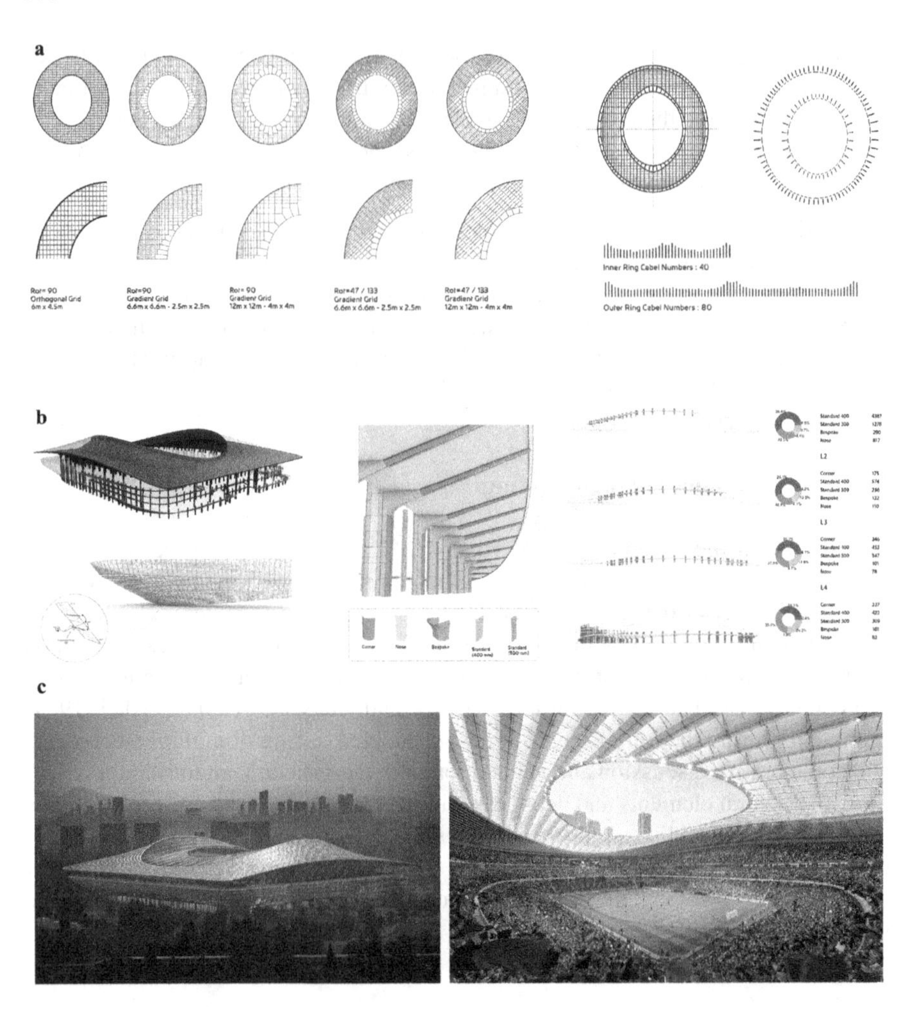

Fig. 13.4 **a** X'ian football stadium, CableNet topological iteration and optimisation. **b** Xian facade optimisation and geometry method statement. **c** X'ian football stadium, Exterior and Interior cable net and bowl renders

amongst collaborators and amenable with game technologies and user engagement with professionally curated content (See Sect. 13.4) (Fig. 13.5).

Critical Constraints of Professionally Generated Content

PGC, despite its design benefits noted in Sect. Benefits of Spatial Technology Stack, is currently expensive to make digitally as the creation of such PGC involves acquisition of considerable digital skills and requires investment in the development of the

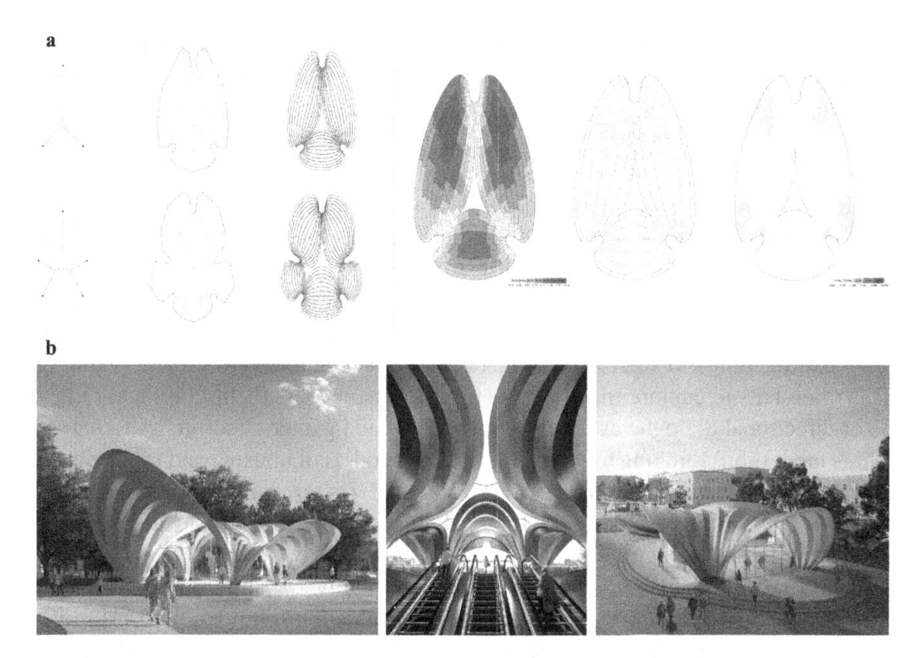

Fig. 13.5 **a** Dnipro Metro Station, Procedural generation of shape from input graph, geometry analysis and rationalisation. **b** Dnipro Metro Station, Exterior and Interior views of the entrance shell canopy

technology stack to create toolsets that are either non-existent or unavailable within commercial design environments. It requires both time and monetary investment in R&D to create and develop sandbox tools which can be field tested via the creation of digital and physical technology demonstrators.

PGC is also currently expensive to make physically as the twentieth century automation-centric production systems are misaligned with STS. Thus, it is currently reliant on RDF and other early-stage technologies and methodologies for its physical realisation.

13.4 User Generated Content and Professionally Generated Content

The motivation to democratise features of AG via web services and game platforms in lieu of PGC is driven by several factors. The desire to deliver mainstream architectural design via a broad base of authorship, effectively crowdsourcing Architecture Engineering and Construction (AEC) practice, has been a long goal. The homebuilders, developers, regulators, and municipalities have looked into it for the purposes of increasing housing stock rapidly, often to the detriment of unit type mix or amenities on offer (Wilson and Barton 2022). The end user is disillusioned in this process resulting in spaces offering utility such as high density yet unfit to serve on cultural criteria such as pedestrian friendly, vibrant, ecological, and economic sustainability (Jacobs 1992; Newman et al. 1996). Likewise, efforts to relinquish centralised regulatory control for instance in housing Permitted Development Rights (PDR), have proved useful optimisations decentralising regulation for specific scope. This included ways to expedite planning permissions, end to end development and engaging of home owners directly (Ministry of Housing 2019).

Democratising co-authorship and participation in design process by non-experts conventionally occurs through the lens of community round tables, focus groups, and town hall meetings to share, disseminate and collate data toward project planning. This often fails to capture the relevant stakeholders or users themselves and fails to be bi-directional dialogue as the design develops and gathers 'inertia.' Alternatives to this model shift thinking toward participatory models of housing and urban design inherited from board games. The participation model is turn based, incentivised, and provides a real time barometer for the 'status' and relative success of the proposal (to those who are playing).

Democratised design poses the distinct opportunity for agile decision making whereby users can change their minds, test and evaluation solutions, shift priorities, arriving at other solutions, and the result is reflective of a changing landscape of decisions (Malmgren et al. 2009; Nahmens and Mullens 2009), not predictable from the outset or deterministic in its resultant form. This offers the distinct advantage of user focussed, consensus driven, market tested solutions which will flourish and be relevant end to end (Hofman et al. 2006; Barlow 1998; Schoenwitz et al. 2012).

Aesthetic and Technical Considerations

The way in which we evaluate STS for UGC is through the 'technical' performance such as polygon budget, limits of computing resources on a device, and the 'social' performance such as user experience, interaction richness, and dramaturgical features of social setting, appearance, and manner of interaction, each contribute to the overall performance of the scheme (Goffman 1956).

There are specific technical and creative challenges to democratising user generated content. The authors of UGC are not likely to have a design background creating the necessity to distribute and encode design 'knowledge' unnecessary in PGC. Technical competencies in software engineering, systems theory, Information Technology (IT) and product development become important as UGC shifts toward STS (Szafir et al. 2016; Moritz et al. 2019; Tang et al. 2011; Langer 1997; Hallstedt, et al. 2020). For instance, UGC is the result of a selection process, not a CAD drawing and creation process in the conventional designerly sense. Not only do a set of selections need to be present, but they also need to be multifarious anticipating a diverse user audience. The creation and management of assets alone introduces workflow challenges of procedural model definition and real time geometry creation. In addition, support for diversity of technical functions including computing resource for physics calculation, rendering, data reconciliation, and synchronisation is needed (Ball 2021).

Further to this is the importance for evaluation and analysis utilities to sort, filter, and return relevant content in the myriad of potential selection objects and corresponding attributes (Nardini et al. 2019). Visual rather than syntactical wayfinding is needed for STS which employs creative capacities in the realm of User Experience (UX) design to develop intuitive and immersive graphical user interfaces (GUI). Through the creation of Augmented Reality (AR) overlays and Heads Up Displays (HUD) product data can be communicated seamlessly, intuitively, and customised to local user profiles and preference (Liu et al. 2010; Chapanis 1959).

Technical considerations include discretisation of system space for digital modularity, as well as preservation of curvilinear shape through discrete voxel representations of content containers.

Trends Toward Industry Alignment of STS

Geometry-based abstraction of complex structural and manufacturing phenomena is an integral feature of AG. This means that many of its core technologies are easily transported to non-expert CAD platforms such as web browsers and game platforms. Gaming platforms are increasingly considered for 'gamified' solutions that require social engagement, multi-stakeholder participation, and negotiation of trade-offs.

The move away from standardised dimensional elements toward mass customisation of assembled components (Sears Roebuck and Co 1936) is supported through innovations in fabrication technology and DfMA (Wood 2021; Thuesen and Hvam 2011). The variability of such coupled with the desire for real time gameplay suggests procedural content generation utilising lightweight inputs as is the case for PGC. Further to this, the creation of in game selections, simulated city fabric and even game level design requires parametric, real-time creation utilising adaptive components for user consumption and design assist creation. This fundamentally makes use of a technology stack for lightweight computation, deployment in series, and field-tested accumulation of knowledge to a common core framework. This suggests a natural extension of end-to-end pipelines to the end user in early design which is technically feasible, and can directly be harnessed for content creation, without an intermediary 'architect.'

Game titles such as Fortnite, Roblox, Second Life, are increasingly de-emphasising goal-oriented gameplay, encouraging the use of the cyber physical for social fulfilment and best practice gameplay mechanics through cooperative play and interactivity (Ross 2014; Maloney 2021; Fabricatore 2007). This coupled with the increasing demand for digital marketplaces for online bidding, negotiation, valuation, collaboration and mechanisms to enhance trust and credibility to potential buyers poses new creative territory where utility for volumetric space alone is not strictly viable (World Economic Forum 2021).

State of UGC

The metaverses are currently too far removed from reality—skirting the boundaries of fantastical, a universe empty by default, operating under its own mechanics. The visual cues are foreign, graphically coarse, or primitive and unintuitive to conventional user wisdom. At the time of publishing, AltspaceVR, Somnium Space, Decentraland each exhibit promising ideas to incentivised building, are navigable, and offer users the opportunity to perform certain utility functions, however they are still non-immersive, interaction sparse, unintuitive, and crude in assimilation to user (Decentraland 2020; Somnium Space 2019; AltspaceVR 2013). This can improve by becoming more familiar by adopting some of the mechanics, 3d-ness, and photo-realism of the physical world to make more intuitive and seamless the experience of switching between online and on-land. This would result in the cyber physical being a precursor-to or as an extension-of the physical experience—an augmented reality not a superseded reality—thereby improving the user experience. We are beginning to see this shift for instance in Non-Fungible Token (NFT) sales by both Christies and Sotheby's in metaspaces as well fashion house exclusive releases for the metaverse by both Gucci and Burberry (Criddle and Klasa 2023).

Benefits of STS in UGC

The following case studies illustrate the benefits of using STS in the design of UGC.

Role of Platform Technology

Platform design initiatives started in 2018 through academic settings at the Architectural Association Design Research Lab (AADRL) and subsequently through ongoing development at ZHA and through workshops, the ongoing studio at AADRL Nahmad-Bhooshan Studio and University College of London (UCL) Bartlett AD RC10. Shifting design thinking to game technologies and platform development at ZHA using Unreal Engine has empowered our capacity to disrupt conventional procurement processes and bring together stakeholders to deliver high value, locally relevant, resource effective, supply chain integrated design solutions.

Platform design helps to *test fit scenarios, explore contingencies, simulate eventualities* to explore the universe of feasible solutions. It facilitates the design of market tested, demand driven solutions delivered and rapidly assembled with more certainty resulting in less risk.

Beyabu Honduras: A Technology Platform and Residential Configurator

The Beyabu Honduras configurator is the latest build in a lineage of residential platforms developed in Unreal Engine offering investors, occupiers and developers ways to position properties onsite to suit, evaluate metrics, and monetise features such as air and development rights in the process. The participants configure modular building components in real time using a web-based application.

The configurator leverages ray tracing to achieve the highest levels of visual fidelity for both the interior and exterior of units. Pixel Streaming enables ZHA to share the configurator experience remotely from the comfort of the participant's browser and device of choice. At the time of publishing the configurator had been used across three different sites exhibiting dramatically different landform characteristics to configure communities.

The platform format brings stakeholders together into a digital marketplace to digitally simulate and negotiate viewpoints in a real time participation platform. This goes beyond the typical product configurator selecting colours, materials, or fixtures in the Ikea-like or real estate spec home catalogue. Users are invited to select from the beyabu residential portfolio of typologies, as well as sequentially position themselves in the community prompted only through certain incentives, including proximity to others, proximity to amenities, total view cone from hill height, as well as real time feedback on selection costs and implications to the aggregate community.

This has demonstrated that AEC industries can engage stakeholders in the design process to co-author, effectively crowd sourcing and democratising the design process. This in part is made possible through a decentralised governance approach in the Honduran Zones for Employment and Economic Development (ZEDE) (Fig. 13.6).

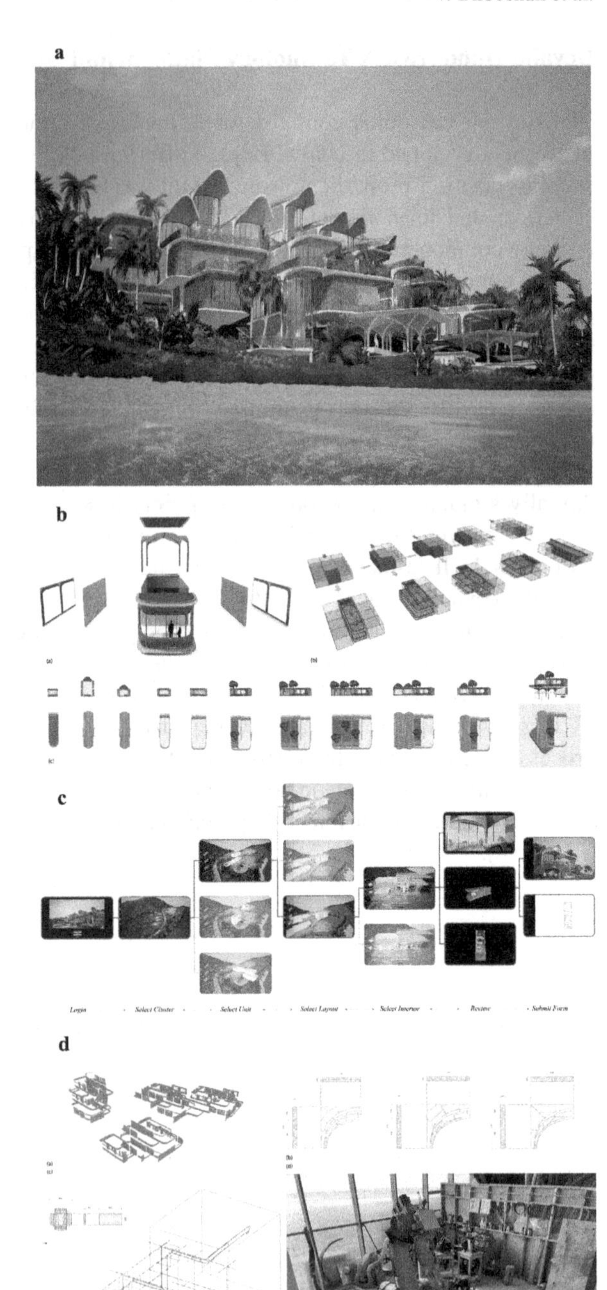

Fig. 13.6 **a** Beyabu configurator—Platform technology and gamification for real estate development. **b** Beyabu configurator—A digital kit of parts encoded to voxels resulting in unit variations. **c** Beyabu configurator—Online user configurator. **d** Beyabu configurator—Robotic assisted digital timber end to end supply chain integration. (**a**) timber cladding elements (**b**) bespoke glulam part creation (**c**) assemblage in relation for discrete spatial representation, (**d**) robot processing of timber element. Image Courtesy of Circular Factory at Hooke Park. **e** Beyabu configurator—Raw user data

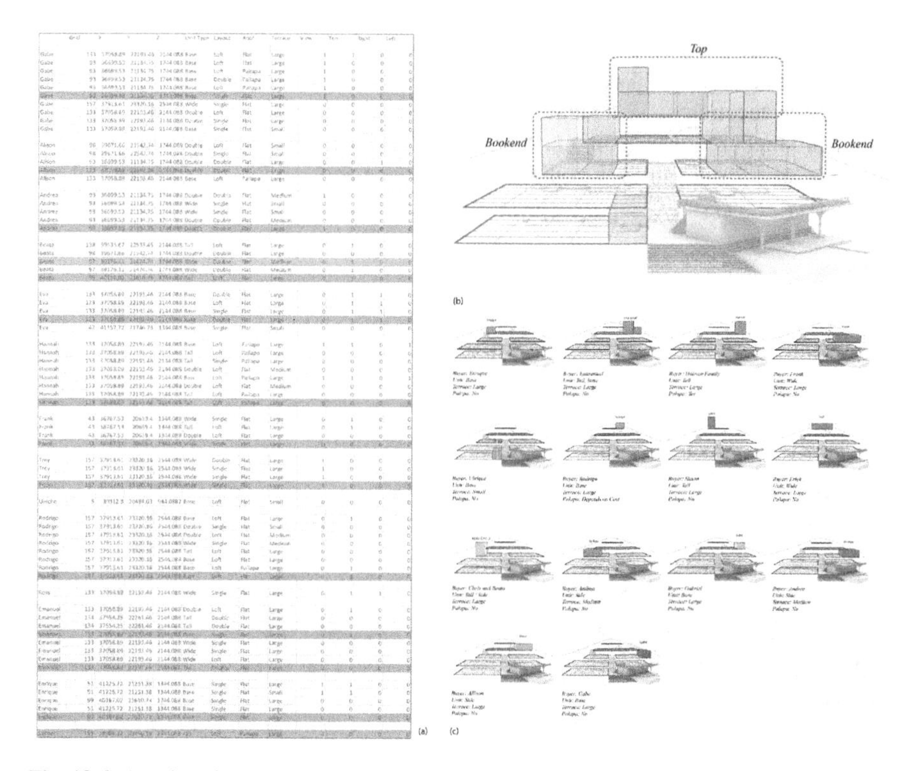

Fig. 13.6 (continued)

Liberland—A Cyber Physical Incubator for Decentralised Finance, Governance, and Urban Planning

The nation of Liberland will first launch through a virtual metaverse to its e-citizens and supporters around the world before its physical launch sometime thereafter. Liberland Metaverse is differentiated from contemporary metaverses for its focus on crypto and the blockchain technology ecosystem in lieu of the entertainment sector. Investment in Liberland Metaverse gains a stake in the physical Liberland. In addition, it is differentiated by its urban and architectural design of the interaction-rich and immersive 3D spatial environment. The urban fabric is characterised by broad open spaces, outdoor activated public spaces, radiating outward from a Central Business District comprised of a series of event venues such as NFT plaza, Decentralised Finance (DeFi) plaza, Exhibition Centre, Incubator, and City Hall. The urban governance model is applicable to both virtual and physical Liberland, leveraging a plurality of planning principles, to offer choice to potential developers, investors, buyers, and end users. These are explored in various districts of Liberland in each sponsored order, self-governed order, and spontaneous order outskirts through a variety of revenue and ownership structures.

Liberland is an attempt to merge the metaverse and web 3.0 vision with urban planning as a decentralised, open participatory model of spatial technologies and governance technologies. It demonstrates there is a mutual relationship between online architecture, urbanism and on-land and as such, can be exploited to enrich, augment, and further fulfil the citizenry experience and economic prosperity of the nation (Fig. 13.7).

a

b

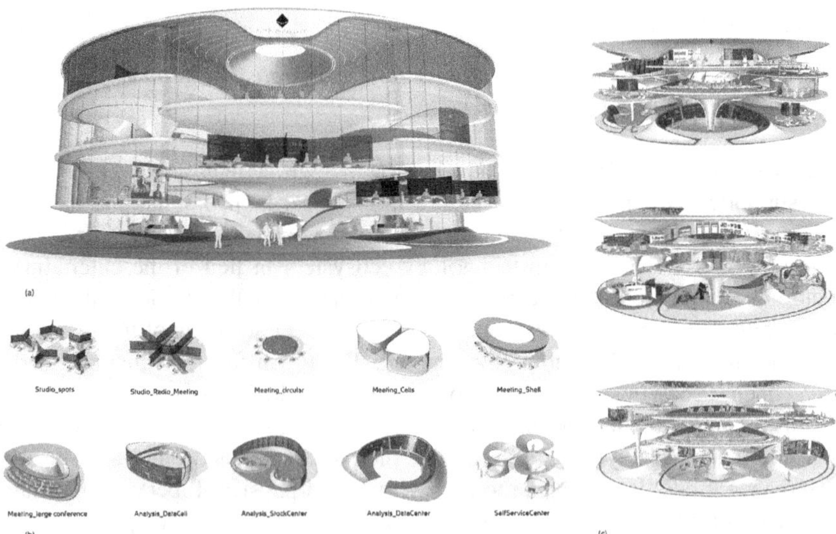

Fig. 13.7 **a** Liberland—a cyber urban incubator. **b** Liberland—(**a**) Incubator Building, (**b**) module variations, (**c**) arrangement variations

Fig. 13.8 **a** PUBG mobile—Erangel Hospital 2050

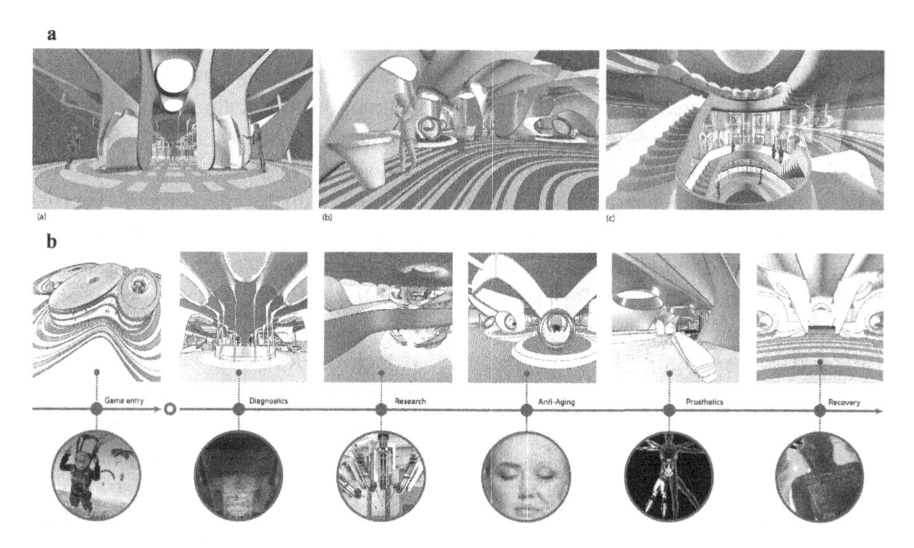

Fig. 13.9 **a** Novel multiplayer battle royale gameplay experiences. **b** Interior differentiation for varied user strategy and tactics and replay value

Player Unknown's Battlegrounds Mobile—Medical Centre

ZHA's joint effort with Player Unknown's Battlegrounds Mobile (PUBG-M) is testament to our deep belief in user-experience (UX) focused design, new technologies of spatial design, novel media of spatial experience, and inter-disciplinary collaboration. The partnership was valuable to reinforce our investment to advance our collaborative, designer-friendly, spatial UX-focussed parametric design technologies.

Spatial designs for the medical centre consider user experience, interaction and navigational aspects that are adapted to 3D shooter game play for each first-person shooter (FPS) and third-person shooter (TPS) scenarios. The online architecture is designed as three interlocking buildings with each one relating to a futuristic hospital use theme—robotic surgery, anti-ageing and longevity research, recuperation and preventative medicines. The furniture and scene entourage were designed in accordance with this creative theme as well as the utility they would serve for sight lines, shelters, and bunkering in gameplay tactics. One of the key design features is the central open atrium space, which whilst providing a clear understanding of how to navigate the building also opened novel vertical combat scenarios, both close and far combat options, and provided players building access via the open rooftop (Figs. 13.8 and 13.9).

13.5 Discussion and Outlook

The discipline of AG is consolidating the research and demonstration gains from its first decade of existence, and progressing towards full scale and mainstream architectural applications with ongoing efforts at the research epicentres in Stuttgart, Zurich and elsewhere (IntCDC 2019; NCCR_dFab 2017; Block et al. 2020). The maturation of several start-up businesses in RDF along with the encoding of expertise in reusable code assets for ease of creation and manipulation of AG, further reinforces this trend of rapid industrialisation of manufacturing and construction technologies (ODICO 2012; AIBuild. AIBuild 2015; BranchTechnology 2015; Jacobson et al. 2016; Mele et al. 2017).

The BIM paradigm, given its documented lack of development, has failed to deliver on its promises of integrated project delivery (IPD), virtual design and construction (VDC) frameworks for delivery of projects (Martyn 2020; Eckblad et al. 2007; Olofsson et al. 2007). This in combination with the difficulty to consider/ represent most discrete geometry representations—predominant information streams in RDF & AG—as building information models (Tolman 1999) making BIM misaligned with the progress in geometry processing, mass customisation. With the paradigm of design for manufacture (DFMA)—using manufacturing input at the earliest stages of the project to design parts that can be produced more easily and more economically (Poli 2001)—taking prominence, STS has the potential to take centre stage as it aligns and is easily embeddable to the software tool chains and collaborative platforms associated with DFMA (Richard 2021; ODICO 2022). Such collaborative platforms aim to incorporate most of the early promises of BIM including:

- increasing engagement of construction knowledge—traditional & RDF—in the design process (Eastman et al. 2011; Khemlani 2009; Sacks et al. 2010);
- developing detailed design earlier than has been common with traditional systems.
- seamless exchange of data and intent among collaborators to reduce time and facilitate iterative refinements (Bhooshan et al. 2018a);

- Increasing flexibility and non-collocated teams (Sacks et al. 2010).

Whilst the deep market moat of BIM helped it to survive the misalignment with RDF & AG, its misalignment with the current production stacks of metaverse and UGC could make it difficult to survive this time around.

The immediate outlook for Spatial Technology Stack (STS) is to significantly improve its prospects of mainstream impact—reducing the costs associated with its digital creation by in turn capturing and encoding the significant tacit know-how that is currently part of the creation process and thus its cost. Such a synergy already underway in the graphics community—Geometric Deep Learning—would help further open the solution space and its exploration, whilst addressing the cost of digitally creating PGC with potential machine assisted creation of Professionally Generated Content (PGC), and would provide a sound basis for disruptive, industry-wide applications of STS in Architecture, Engineering and Construction.

References

Akbarzadeh M, Van Mele T, Block P (2015) On the equilibrium of funicular polyhedral frames and convex polyhedral force diagrams. Comput-Aided Des 63:118–128

AIBuild. AIBuild (2015)

AltspaceVR (2013) AltspaceVR. https://altvr.com/. Accessed 18 Feb 2022

Ball M (2021) The metaverse primer. https://www.matthewball.vc/the-metaverse-primer. Accessed 18 Feb 2022

Barlow J (1998) From craft production to mass customization? Customer focused approaches to house building. In: Proceedings international group for lean construction IGLC

Bhooshan S, El Sayed M (2011) Use of sub-division surfaces in architectural form-finding and procedural modelling. In: Proceedings of the 2011 symposium on simulation for architecture and urban design, pp 60–67

Bhooshan S, Bhooshan V, ElSayed M, Chandra S, Richens P, Shepherd P (2015) Applying dynamic relaxation techniques to form-find and manufacture curve-crease folded panels. Simulation 91(9):773–786

Bhooshan S (2016a) Collaborative design—A case for combining CA(G)D and BIM. Archit Des

Bhooshan S (2016b) Upgrading computational design. Archit Des 86

Bhooshan S (2016c) Realizing architecture's disruptive potential. Oz 38(1):7

Bhooshan S (2017) Parametric design-thinking A case-study of practice-embedded architectural research. Des Stud

Bhooshan S, Bhooshan V, Dell'Endice A, Chu J, Singer P, Megens J, Block P (2022a) The Striatus bridge: computational design and robotic fabrication of an unreinforced, 3D-concrete-printed, masonry arch bridge. Archit Struct Constr 1–23

Bhooshan S, Bhooshan V, Megens J, Casucci T, Van Mele T, Block P (2022b, September) Print-path design for inclined-plane robotic 3D printing of unreinforced concrete. In: towards radical regeneration: design modelling symposium Berlin 2022. Springer International Publishing, Cham, pp 188–197

Bhooshan V, David Louth H, Bhooshan S, Schumacher P (2018a) Design workflow for additive manufacturing: a comparative study. Int J Rapid Manuf 7(2–3):240–276

Bhooshan S, Ladinig J, Van Mele T, Block P (2018b) Function representation for robotic 3D printed concrete, ROBARCH 2018b—Robotic fabrication in architecture, art, and design. SpringerZurich, pp 98–109

Bhooshan S, Van Mele T, Block P (2018c) Equilibrium-aware shape design for concrete printing. In: Humanizing digital reality. Springer, Singapore, pp 493–508
Bhooshan V, Reeves D, Bhooshan S, Block P (2018d) MayaVault—a mesh modelling environment for discrete funicular structures. Nexus Netw J 20(3):567–582
Bhooshan V, Louth H, Bieling L, Bhooshan S (2019) Spatial developable meshes. In: Design modelling symposium Berlin. Springer, Cham, pp 45–58
Bhooshan S, Vazquez AN (2020) Homes, communities and games: constructing social agency in our urban futures. Archit Des 90:60–65
Block P (2016) Parametricism's structural congeniality. Archit Des 86:68–75
Block P, Ochsendorf J (2007) Thrust network analysis: a new methodology for three-dimensional equilibrium. J Int Assoc Shell Spat Struct 48(3):167–173
Block P, Van Mele T, Rippmann M, Ranaudo F, Javier Calvo Barentin C, Paulson N (2020) Redefining structural art: strategies, necessities and opportunities. Struct Eng 98(1):66–72
BranchTechnology. Branch technology (2015). https://www.branch.technology/. Accessed 12 July 2019
Bruns A (2007) The future is user-led: the path towards widespread produsage. In Proceedings of perthDAC 2007: the 7th international digital arts and culture conference. Curtin University of Technology, pp. 68–77
Chapanis A (1959) Research techniques in human engineering. John Hopkins University Press, Baltimore, MA
Criddle C, Klasa A (2023) What Gucci and others learnt from the metaverse. https://www.ft.com/content/d4c3d51f-4568-400e-8ca9-7706539d9cae. Accessed 24 Feb 2023
Decentraland (2020) Decentraland. https://decentraland.org/. Accessed 18 Feb 2022
Eastman CM, Eastman C, Teicholz P, Sacks R, Liston K (2011) BIM handbook: a guide to building information modeling for owners, managers, designers, engineers and contractors. John Wiley & Sons
Eckblad S, Ashcraft H, Audsley P, Blieman D, Bedrick J, Brewis C, Hartung RJ, Onuma K, Rubel Z, Stephens ND (2007) Integrated project delivery-a working definition. AIA California Council, Sacramento, CA 25
Fallacara G (2009) Toward a stereotomic design: experimental constructions and didactic experiences. In: Proceedings of the third international congress on construction history 553
Fabricatore C (2007) Gameplay and game mechanics design: a key to quality in videogames. https://doi.org/10.13140/RG.2.1.1125.4167
Goffman E (1956) The presentation of self in everyday life. University of Edinburgh Social Sciences Research Centre
Hallstedt S et al (2020) The need for new product development capabilities from digitalization, sustainability, and servitization trends. Sustainability 12, 10222. https://doi.org/10.3390/su1223 10222
Hofman E, Halman JI, Ion RA (2006) Variation in housing design: identifying customer preferences. Hous Stud 21:929–943. https://doi.org/10.1080/02673030600917842
IntCDC (2019) Cluster of excellence integrative computational design and construction for architecture. https://icd.uni-stuttgart.de/?p=24111
Jacobs J (1992). The death and life of great American cities
Jacobson A et al (2016) libigl: A simple C++ geometry processing library. 2018-10-16 [2019-06-03]. https://libigl.github.io
Jenkins H (2006) Fans, bloggers, and gamers: exploring participatory culture. Nyu Press
Jiang C, Tang C, Tomičí M, Wallner J, Pottmann H (2015) Interactive modeling of architectural freeform structures: combining geometry with fabrication and statics. In: Advances in architectural geometry 2014. Springer, Cham, pp 95–108
Khemlani L (2009) Sutter medical center castro valley: case study of an IPD project. AECBytes, Khemlani L, CA. http://www.aecbytes.com/buildingthefuture/2009/Sutter_IPDCaseStudy.html
Lee J (2018) Computational design framework for 3D graphic statics. PhD diss., ETH Zurich

Louth H, Reeves D, Koren B, Bhooshan S, Schumacher P (2017) A prefabricated dining pavilion: using structural skeletons, developable offset meshes, kerf-cut and bent sheet materials. Fabricate:58–67

Lobato R, Thomas J, Hunter D (2012) Histories of user-generated content: between formal and informal media economies. In: Amateur media. Routledge, pp 19–33

Liu Y, Osvalder A-L, Karlsson M (2010) Considering the importance of user profiles in interface design. https://doi.org/10.5772/8903

Langer AM (1997) Analysis and design of information systems (3rd ed)

Malmgren L, Jensen P, Olofsson T (2009) Product modelling of configurable building systems—A case study. Submitted December 2009 to Journal of Information Technology in Construction

Maloney D (2021) A youthful metaverse: towards designing safe, equitable, and emotionally fulfilling social virtual reality spaces for younger users. All Dissertations. 2931

Martyn D (2020) The future of Revit. https://aecmag.com/bim/the-future-of-revit/. Accessed 20 Feb 2022

Michalatos P, Payne A (2016) Monolith: the biomedical paradigm and the inner complexity of hierarchical material design

Ministry of Housing, Communities, and Local Government. (2019). https://assets.publishing.service.gov.uk/government/uploads/system/uploads/attachment_data/file/830643/190910_Tech_Guide_for_publishing.pdf. Accessed 18 Feb 2022

Moritz D et al (2019) Formalizing visualization design knowledge as constraints: actionable and extensible models in Draco. IEEE Trans Visual Comput Graph 25(1):438–448. https://doi.org/10.1109/TVCG.2018.2865240

Nahmens I, Mullens M (2009) The impact of product choice on lean homebuilding. Constr Innov 9:84–100. https://doi.org/10.1108/14714170910931561

Nardini FM, Trani R, Venturini R (2019) Fast approximate filtering of search results sorted by attribute. 815–824. https://doi.org/10.1145/3331184.3331227

NCCR_dFab (2017) National centre of competence in research (NCCR) digital fabrication. http://www.dfab.ch/

Newman O, Rutgers University, United States (1996) Creating defensible space. U.S. Dept. of Housing and Urban Development, Office of Policy Development and Research, Washington, DC

ODICO (2012) Odico robotic technologies. https://www.odico.dk/en/technologies. Accessed 20 July 2019

ODICO (2022) Factory on the fly. https://odico.dk/en/factoryonthefly/. Accessed 20 Feb 2022

Olofsson T, Lee G, Eastman C, Reed D (2007) Benefits and lessons learned of implementing building virtual design and construction (VDC) technologies for coordination of mechanical, electrical, and plumbing

Poli C (2001) Design for manufacturing: a structured approach. Butterworth-Heinemann

Pottmann H (2010) Architectural geometry as design knowledge. Archit Des 80:72–77

Panozzo D, Block P, Sorkine-Hornung O (2013) Designing unreinforced masonry models. ACM Trans Graph (TOG) 32(4):1–12

Pottmann H, Brell-Cokcan S, Wallner J (2006) Discrete surfaces for architectural design. Curves Surf: Avignon 213:e234

Prévost R, Whiting E, Lefebvre S, Sorkine-Hornung O (2013) Make it stand : balancing shapes for 3D fabrication. ACM Trans Graph (TOG) 32(4):1–10

Rabinovich M, Hoffmann T, Sorkine-Hornung O (2018) The shape space of discrete orthogonal geodesic nets. ACM Trans Graph (TOG) 37(6):1–17

Reeves D, Bhooshan V, Bhooshan S (2016) Freeform developable spatial structures. In: Proceedings of IASS annual symposia, vol 2016, no 3, pp 1–10. International Association for Shell and Spatial Structures (IASS)

Richard H (2021) Design for constructability. https://aecmag.com/opinion/design-for-constructability/. Accessed 20 Feb 2022

Rippmann M, Liew A, Van Mele T, Block P (2018) Design, fabrication and testing of discrete 3D sand-printed floor prototypes. Mater Today Commun 15:254–259

Ross B (2014) General video game playing with goal orientation. Thesis
Sacks R, Koskela L, Dave BA, Owen R (2010) Interaction of lean and building information modeling in construction. J Constr Eng Manag 136(9):968–980
Schek H-J (1974) The force density method for form finding and computation of general networks. Comput Methods Appl Mech Eng 3(1):115–134
Schumacher P (2014a) Tectonic articulation: making engineering logics speak. Archit Des 84:44–51
Schumacher P (2014b) The congeniality of architecture and engineering. Shell Struct Archit Form Find Optim 271
Schoenwitz M, Naim M, Potter A (2012) The nature of choice in mass customized house building. Constr Manag Econ 30:203–219. https://doi.org/10.1080/01446193.2012.664277
Schwartzburg Y, Pauly M (2013) Fabrication-aware design with intersecting planar pieces. In: Computer graphics forum, vol 32, no 2pt3. Blackwell Publishing Ltd., Oxford, UK, pp 317–326
Sears Roebuck and Co (1936) Modern homes. https://dahp.wa.gov/sites/default/files/ModernHomessears1936.small_.pdf. Accessed 4 Dec 2021
Somnium Space (2019). Somnium Space. https://somniumspace.com/. Accessed 18 Feb 2022
Szafir DA, Haroz S, Gleicher M, Franconeri S (2016) Four types of ensemble coding in data visualizations. J Vis 16(5):11. https://doi.org/10.1167/16.5.11
Tamke M (2015) Aware design models. In SpringSim (SimAUD), pp 213–220
Tang A, Gerrits T, Nacken P, Vliet H (2011) On the responsibilities of software architects and software engineers in an agile environment: who should do what? In: SSE'11—Proceedings of the 4th international workshop on social software engineering. https://doi.org/10.1145/2024645.2024650
Tang C, Sun X, Gomes A, Wallner J, Pottmann H (2014) Form-finding with polyhedral meshes made simple. ACM Trans Graph (TOG) 33(4):1–9
Thuesen CL, Hvam L (2011) Efficient on-site construction: learning points from a German platform for housing. Constr Innov 11(3):338–355. https://doi.org/10.1108/14714171111149043
Tolman FP (1999) Product modeling standards for the building and construction industry: past, present and future. Autom Constr 8(3):227–235
Van Mele T, Liew A, Mendez T, Rippmann M (2017) COMPAS: a framework for computational research in architecture and structures
Vouga E, Höbinger M, Wallner J, Pottmann H (2012) Design of self-supporting surfaces. ACM Trans Graph (TOG) 31(4):1–11
Wilson W, Barton C (2022) Tackling the Under-Supply of Housing. House of Commons Library. Number CBP-7671. https://researchbriefings.files.parliament.uk/documents/CBP-7671/CBP-7671.pdf. Accessed 18 Feb 2022
Wood B (2021) Delivery platforms for government assets creating a marketplace for manufactured spaces. https://www.brydenwood.co.uk/platformdesignbooks/s114123/. Accessed 4 Dec 2021
World Economic Forum (2021) Next-generation business models a guide to digital marketplaces. https://www3.weforum.org/docs/WEF_Marketplaces_guidebook_2021.pdf. Accessed 18 Feb 2022

Chapter 14
Why Disruptive Business Models are Inseparable from Disruptive Technologies

Philippe Morel

This chapter on the relationship between business models and disruptive technologies can be seen as a counterpoint to the general theme of this volume "*Disruptive Technologies: The Convergence of New Paradigms in Architecture*". The latter insists on the technologies themselves, most often from the point of view of their technical operativity or from an epistemological perspective. We will see here that beyond the operativity of the technology itself and the epistemology associated with technology, which creates new conceptual frameworks of understanding, the insertion of technologies into the global economic market—which publicizes everything and through which everything passes—plays an ever more important role. This insertion obviously impacts the technological appropriation and the technological evolution itself, including in architecture perceived here in a broad sense, from the conception to the maintenance of projects after delivery.

If current architecture is increasingly defined in relation to the incessant technological evolutions (hardware and software) that it integrates into the different stages of a project, from design to handover to the client, it would be a mistake to believe, on the one hand, in a technological determinism that is solely responsible for the evolution of this discipline, and, on the other hand, in the complete autonomy of the technology. As Clayton Christensen, the inventor of the notion of *disruptive innovation* (Bower and Christensen 1995), has shown—and as the simple fact that Christensen is not a technologist, engineer, or inventor but an economist and consultant shows—a technology only becomes truly interesting when it moves from the status of disruptive technology to the more fundamental status of *disruptive innovation*. Indeed, whereas a disruptive technology, in the current *but inaccurate* sense

P. Morel (✉)
The Bartlett School of Architecture, Faculty of the Built Environment, University College London (UCL), London, UK
e-mail: p.morel@ucl.ac.uk

P. Morel and H. Bier (eds.), *Disruptive Technologies: The Convergence of New Paradigms in Architecture*, Springer Series in Adaptive Environments,
https://doi.org/10.1007/978-3-031-14160-7_14

of the term, can be limited to merely being a "new", "radical", "futuristic" technology, and sometimes all of these at the same time, with no guarantee of success (cf. the immense number of inventions that have never met with commercial success), disruptive innovation is intrinsically associated with *market penetration.* It is even its definition. Disruptive innovation operates through the creation of a new market and a value-adding network, or through the entry into an existing market in which the positions of the players are contested and displaced. As such, a disruptive technology is not a guarantee of success, it could even be a guarantee of failure if we follow what J.L. Bower and C. Christensen claimed in their 1995[1] programmatic text: "*On the other hand,* disruptive *technologies introduce a very different package of attributes from the one mainstream customers historically value, and often perform far worse along one or two dimensions that are particularly important to those customers.*"[2] It is therefore not its performance and qualities that make a disruptive technology interesting, since "*as a rule, mainstream customers are unwilling to use a disruptive product in applications they know and understand*", but rather its potential, its ability to create new uses and therefore new markets: "*At first, then, disruptive technologies tend to be used and valued only in new markets or new applications; in fact, they generally make possible the emergence of new markets. For example, Sony's early transistor radios sacrificed sound fidelity but created a market for portable radios by offering a new and different package of attributes—small size, light weight, and portability.* " If we look at what is happening in architecture around issues of technological innovation and more specifically disruptive technology, we find that most architects and inventors active in this field too often confuse radicality and disruption about technology. As V. Govindarajan and P.K. Kopalle pointed out, "*disruptive innovations can be high end as well, i.e., technologically more radical in nature*", but more importantly "*the disruptiveness construct is distinct from the radicalness dimension. The* radicalness *of innovations refers to the extent an innovation is based on a substantially new technology relative to existing practice [...]. On the other hand, the* disruptiveness *of innovations refers to the extent an emerging customer segment, and not the mainstream customer segment, sees value in the innovation at the time of introduction, which, over time, disrupts the products mainstream customers use [...].*" In reality, and this is the main point, "*the radicalness is a technology-based dimension of innovations, and the disruptiveness is a market-based dimension* (Govindarajan and Kopalle (2006))." This distinction between the commercial and technical dimensions of technological innovation was made by C. Christensen himself, in *The Innovator's Solution: Creating and Sustaining Successful Growth* (Christensen and Raynor 2003), a book he co-authored with M. Raynor in response to his famous, and first book, *The Innovator's Dilemma* (Christensen 1997) of 1997. In this book, the authors distinguish several types of disruptions: low-end disruptions and high-end disruptions. Low-end disruptions are characterized by a market penetration based on products that are less efficient (or in rare cases with performances similar to the existing ones) but also less expensive, while high-end disruptions offer

[1] Bower and Christensen (1995)

[2] Ibid., p.45.

new products and new services. Although the radicality of technical innovation as such is not always exacerbated and equally distributed, it nevertheless plays a major role in the latter type of disruptive innovation. However, for Christensen, the problem of technological disruption remains above all a marketing problem and not a technological problem—which the history of technological innovation tends to confirm. Hence, for a company founder, a disruptive strategy must be understood from a marketing perspective, otherwise the products may not find or create a market, and this is also the perspective from which an economist must study it: "*I think this is a reason why my research was able to add value [...]. I examined the phenomena through the lenses of marketing and finance and not just the technological dimensions of the problem, which allowed me to see things that others had not seen before* (Christensen 2006). "As for how one can identify the disruptiveness of innovation, this is a difficult question. If radical (high-end) technological innovation attracts people's attention, it seems that for low-end innovation the task is far more difficult. It even seems that the "*measure of disruptiveness is indeed ex post; that is, one can assess the disruptiveness of an innovation only after it has been introduced and begins to disrupt the mainstream market.*"[3]

As mentioned, disruption is generally perceived in architecture from the point of view of technological radicalism, which is constantly demonstrated by the projects carried out in schools of architecture. But this radicality is not a guarantee of success, far from it. It even tends to challenge the status quo too head-on, creating a handicap that the early adopters market, which is necessary for the success of a disruption, is not always enough to overcome. There are few disciplines in which, like architecture, the intelligence generated in schools and universities has so little concrete and visible effect on the market and in everyday life. One reason for this is that the market is almost always ignored. Although there are rare exceptions that come close, such as Phil Bernstein's *Exploring New Value Propositions of Design Practice* at Yale University (Bernstein 2020), there are no economics modules in the curricula of future architects,[4] for example. The understanding by students, but also by teachers and practitioners, of the expectations of consumers, whoever they may be, is too low.[5] This is due to an even weaker understanding of the market and the economy, although the Greek root of the term—*Oikonomia*—associates the notion of home with that of good management. We might counter that many practitioners understand the expectations of clients on a *daily* basis. This is true, but it is not comparable to the understanding that allows the development of new business models that enable architecture in the broadest sense, or architects as key players, to at least adapt to the global evolution of the market and at most to modify—even if only slightly—the trajectory of this evolution. Whether an architect chooses a low-end

[3] Govindarajan & Kopalle (2006), *Ibid.*

[4] The courses associated with obtaining licenses to practice, such as in France the HMONP (habilitation à la maîtrise d'oeuvre en nom propre) partially deal with the economics of architecture firms, but in no case can these courses replace economics courses that could be given at the undergraduate level.

[5] It goes without saying that the participation of inhabitants in the design process, a fashionable "solution" in the late 60 s and early 70 s, obviously does not equate to any kind of deep understanding of what the market needs.

or high-end disruption strategy is of little importance, at least much less so than being able to choose one of them and apply it thanks to an innovative business model. The need for such models is manifold. They are necessary for architects to extract themselves from a remuneration model that appeared at the end of the Middle Ages, a model that is obsolete in a society of information, networks, and artificial intelligence, whose "original paradox" Ph. Bernstein (among many practicing architects) has rightly noted: *the more the architect works, the less he/she is paid*. The main consequence of this low remuneration is—at least for me—not the inability to get rich, but the impossibility to invest and to do what every company today needs, namely research and development (R&D). As Bernstein notes, referring to George Barnett Johnston's book[6] and the fictional character from early twentieth-century literature he invokes, Tom Thumtack: "*Tom was explaining the seeming illogic of a system of compensation for architects that had, within it, two deep contradictions. First, when the architect's fee is based on a percentage of construction cost, the harder the architect works to bring the project into cost conformance, the less she is paid. Second in the cases so common today when said fee is converted into a lump sum, the client has transferred the financial risk of the fee over the architect, who perversely is now incentivized to work less, rather than more, to service that client, and thereby preserve some remainder of the fee as profit*." Beyond this widely shared remuneration issue, understanding the intertwining of disruptive technologies and disruptive business models is a more crucial problem in the face of current challenges, for example those related to the need to build massively without making the same mistakes as in the past.[7] To do so, architects must expand their field of influence as much as their knowledge. Indeed, while more and more professions intervene in the field of architecture and construction, and while more and more tools—e.g. either user-friendly 3D design tools such as SketchUp® or advanced simulation software, all of which are enrichable by AI—are profoundly modifying the sociology of the discipline, architects are restricted to a traditionally circumscribed practice. This practice does not differ from the type of practice most publicized in the heroic modernism of the 1920s and 1930s, whereas this same modernism, in reality anything but monolithic, was full of entrepreneurial innovations on the part of architects. If, as for the pioneers of modernism, to envisage the future with prescience remains necessary, it guarantees nothing. The technique may not be available or may not be sufficiently developed, as was the case when McLuhan declared, in 1968, that "*When electronically controlled devices are perfected, it will be almost as simple and cheap to obtain a million different objects as to make a million identical ones*." McLuhan was anticipating what would later be called non-standard production, but, on the one hand, this idea was not relevant at the time when the mass consumer society (of standard products) was reaching its peak, and, on the other hand, McLuhan was not interested in developing a business… In which case, as an entrepreneur, he could perhaps have led the development of suitable machines and

---

[6] See Johnson (2020). Johnston quotes Squires and Kent (1914).

[7] Those of post-war modernism in the United States, England and France, mainly under the disastrous theoretical influence of Team X and structuralism.

business models. The vision of a future fundamentally articulated around a technique may not be shared, even by the most knowledgeable people. The famous example of Ken Olsen, the founder of Digital Equipment Corporation (DEC), stating in 1977 that "*There is no reason for any individual to have a computer in his home*" is emblematic of the difficulty of envisaging the future of a technology.[8] This is all the more true since the evolution of the current society depends on a number of parameters infinitely superior to what we were used to in the past, which motivated the application of complexity theories in the social and economic field, and which today motivates the massive use of AI techniques capable of dealing with millions of parameters.

If we go back to architecture in the strict sense, what would be the most obvious disruptive technologies today? On the software side, certainly the artificial intelligence techniques discussed in this volume, AI applied to design, project management, site management, and product lifecycle management (PLM). On the hardware side, everything that concerns robotics and that is also covered in this volume, and of course 3D printing in the broadest sense. As the initiator, founding CEO, and business angel of a company (XtreeE) active in this latter field, it goes without saying that this is an area I know particularly well. It is largely because I have been studying it for many years that I have been able to create a solid startup that is not among the nine out of ten startups that fail (one can also apply Warren Buffett's famous tip—"*invest in businesses that are so wonderful that an idiot can run them. Because sooner or later, one will*"—, which I maybe did unconsciously...). But to say that the success of such a company is a certainty even before its legal creation would be presumptuous. To know for sure whether a disruptive technology, defined precisely—for example a very specific 3D printing technique with materials that are no less specific—or defined generically—for example AI techniques—will be successful is of course impossible. There is no magic recipe for success, whatever the qualities of the technology developed. Nevertheless, some economic fundamentals remain. Thus, the first reason why a startup fails is simply that it offers products that nobody wants... "*A careful survey of failed startups determined that 42% of them identified the 'lack of a market need for their product' as the single biggest reason for their failure* (Griffith 2014)." The second is obviously to ignore the reality of business in favor of innovation alone, technological innovation or business model innovation, or the only parameters that we spontaneously associate with creative work. As one founder said, "*a good product idea and a strong technical team are not a guarantee of a sustainable business. One should not ignore the business process and issues of a company because it is not their job. It can eventually deprive them from any future in that company* (Parisot 2022)." Another reason is related to a poor understanding of the chains of market players, the mechanisms of competition or exclusivity, or the

[8] Olsen later explained that the quote was taken out of context and that he was referring not to PCs but to computers set up to control houses. If we consider this latter option, we could say that he was also wrong. In any case, this example does not mean that such persons lack capacity of anticipation. It mostly shows that considering the right set of parameters influencing the future is a challenging task. We should also keep in mind that other technological inventions could have led to highly different paths and histories in which Olsen's prediction would prove right.

regulatory environment and issues related to intellectual property. On this last point, it is worth remembering that architecture remains a fundamentally intellectual discipline organized around the notion of intellectual property, even if architects are not always aware of it on a daily basis. Intellectual property is a legal protection granted to a producer of original knowledge. It allows this same producer to monetize the exchange of the knowledge he has produced. But what is this knowledge, or rather what are the exact limits, when more and more software tools are involved in the production process? There is no simple answer to this question. While it is fairly obvious that 2D or 3D models created "from scratch" in standard modeling software belong to their creators, the same cannot be said for the use of certain artificial intelligence "software". If Dalle·E, created by OpenAI, grants users the right to use the generated images (called "Generations") as they wish, including by selling their rights ("*you may use Generations for any legal purpose, including for commercial use. This means you may sell your rights to the Generations you create, incorporate them into works such as books, websites, and presentations, and otherwise commercialize them.*"), the ownership of the same images does not actually belong to the users. They own the "Prompts" and "Uploads" (used to generate images) but not the generated images themselves (the "Generations"). As mentioned in article "*6. Ownership of Generations*" of the Dalle·E[9] terms of use, "*To the extent allowed by law and as between [you] and OpenAI, [you] own [your] Prompts and Uploads, and [you] agree that OpenAI owns all Generations (including Generations with Uploads but not the Uploads themselves), and [you] hereby make any necessary assignments for this.*" This remains relatively traditional since immediately after this reminder an exclusive right of use is granted to the user (should we say "naturally"): "*OpenAI grants you the exclusive rights to reproduce and display such Generations and will not resell Generations that [you] have created, or assert any copyright in such Generations against [you] or [your] end users, all provided that [you] comply with these terms and our Content Policy.*" The situation becomes more complex, however, when rights similar to those granted to user A are granted to other users, B, C, etc.: "*[you; i.e. user A] understands and acknowledges that similar or identical Generations may be created by other people [i.e. users B, C,* etc.*] using their own Prompts, and [your; i.e. user A] rights are only to the specific Generation that [you; i.e. user A] have created.*" From the point of view of traditional architectural creation, what interests the potential client of an architect (or artist) is not the way in which he or she would have generated the result—although this is for the architect inseparable from the result—but the result itself. What happens then if a designer edits the input text (the so-called Prompt) in such a way that the result is absolutely identical to a pre-existing result? What would traditionally—and quite easily—be considered plagiarism does not seem to fall into this category here. What we should retain from this basic example, which is related to architecture but which does not seem to be directly associated with the broader issue of intellectual property in business, or for example more specifically in technology startups, is that beyond the

[9] Https://labs.openai.com/policies/terms.

difficulty of considering the evolution of technologies for their own sake, the difficulty of considering the legal and normative consequences is no less great. What some might consider a purely legal problem is in fact one of the many parameters that come into play in the development of business models, especially when these models are intended to be disruptive. From these various remarks, we can conclude that separating disruptive business models from disruptive technologies is a short view. We should therefore regret that the discourse on technology in architecture does not pay enough attention to the complexity of the transition between an idea and its success in the real world. While anticipating needs that do not yet exist is naturally associated with any disruptive approach, understanding existing needs and forces should not be dismissed.

References

Bernstein P (2020) The distractions of disruptions: technical supply in an era of social demand. Arch. Des. (AD) 90(02)

Bower JL, Christensen CM (1995) Disruptive technologies: catching the wave. Harv Bus Rev 73:43–53

Christensen CM (1997) The innovator's dilemma: when new technologies cause great firms to fail. Harvard Business School Press, Boston, Massachusetts, USA

Christensen CM (2006) The ongoing process of building a theory of disruption. J Prod Innov Manag 23:39–55

Christensen CM, Raynor EM (2003) The innovator's solution: creating and sustaining successful growth. Harvard Business Press, Cambridge, MA

Govindarajan V, Kopalle PK (2006) The usefulness of measuring disruptiveness of innovations ex post in making ex ante predictions. J Prod Innov Manag 23:12–18

Griffith E (2014) Why startups fail, according to their founders. Fortune magazine

Johnson GB (2020) Assembling the architect: the history and theory of professional practice. Bloomsbury Visual Arts, London

Parisot T, https://thom4.net/2014/why-our-startup-failed. last access February 2022

Squires F, Kent R (1914) Architec-tonics: The Tales of Tom Thumtack, Architect. The William T. Comstock Company, New York

Printed in the United States
by Baker & Taylor Publisher Services